Unit 28

Geometry in the Workplace 1

Developed by the Center for Occupational Research and Development and sponsored by a consortium of State Vocational Education Agencies with the cooperation and support of mathematics educators.

PREFACE

Applied Mathematics contains video programs, laboratory activities and problem-solving exercises. Each part has been chosen to help you understand the mathematics you need to work and live in a technical world. Most importantly, each part has been designed to make mathematics more useful and meaningful for you—and reduce some of the "math anxiety" we all feel at one time or another.

Applied Mathematics contains some of the important mathematics you'll need to be a productive member of today's workforce. The ideas you'll learn will help you understand:

- numbers, decimals, fractions and percents;
- shapes and sizes;
- how to handle equations and formulas;
- how to work with angles and triangles;
- how to estimate answers and solve problems;
- how to describe the behavior of large populations of things;
- how to use computers in problem solving;
- how to use scientific and graphing calculators to solve problems.

Each of these skills will help you do your job better—and help you advance to the next job when the time comes.

Each unit of **Applied Mathematics** begins with a *video program*. The video tells you about the mathematics skills you'll be studying and introduces you to real people who use these skills in their everyday lives. Following this, you'll concentrate on learning the *mathematics skills*.

Next, you'll have a chance to practice what you've learned. You'll do this in *laboratory activities* that involve measurement and in *problem-solving exercises*. You'll apply the math skills to practical problems. These problems are the kind that people have to solve every day—in restaurants, on the farm, in a factory, in a business office, in a hospital, in a laboratory, or at home.

You'll have many opportunities to practice what you've learned, so don't be discouraged if it all doesn't seem too clear the first time around. You'll learn how to use mathematics, little by little, problem by problem. And in the process, you will find that using mathematics can be fun.

The CORD Project Staff

Table of Contents

Geometry in the Workplace 1

How to apply the principles of geometry to solve problems in the workplace

Prerequisites

This unit builds on the skills taught in
Unit 2: *Estimating Answers*
Unit 3: *Measuring in English and Metric Units*
Unit 6: *Working with Lines and Angles*
Unit 7: *Working with Shapes in Two Dimensions*
Unit 10: *Working with Scale Drawings*
Unit 12: *Using Scientific Notation*
Unit 13: *Precision, Accuracy and Tolerance*
Unit 15: *Using Formulas to Solve Problems*
Unit 21: *Using Right-triangle Relationships*
Unit 22: *Using Trigonometric Functions*
Unit 23: *Factoring*
Unit 25: *Quadratics*

To Master This Unit

Read the text and answer all questions. Complete the assigned activities and exercises. Work the problems on the unit test at a satisfactory level.

Unit Objectives

Working through this unit helps you learn how to:

1. Apply geometry to problems that involve the areas of rectangles, triangles, trapezoids, circles, sectors and segments of circles.

2. Apply geometry to problems that involve tangent lines, perpendicular lines, bisecting lines and angles.

3. Apply the Pythagorean theorem and the sine, cosine and tangent functions to right triangles to relate sides and angles.

4. Draw auxiliary diagrams to help solve for an unknown dimension or an unknown angle.

5. Solve geometry problems that involve a series of successive calculations.

1. Read the "Introduction" section of this unit.

2. Watch the video and take part in the class discussions.

3. Read and study the text. Study the examples closely.

4. Do the assigned lab activity.

5. Complete the assigned exercises.

6. Measure your progress by taking the unit test.

Some Signals to Help You Learn

The following signals help you know what to do as you read the text:

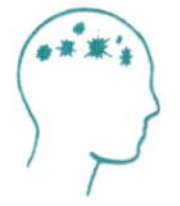 Think this through. Spend a little extra time on this idea.

 Write your answer on your paper.

 Carry out the calculations.

 Learn this key rule or definition.

 Estimate and ask yourself if this answer makes sense.

 Compare your answer to the given one and make any needed changes.

INTRODUCTION

Geometry is an important part of everyday life. Virtually every occupation uses the principles of geometry in one form or another. The Egyptians found geometry to be useful when they built the pyramids thousands of years ago. Today, in the space age, geometry is even more useful.

In this unit, you will learn more about applications of geometry in the workplace. Whether the workplace happens to be your home, the farm, the office, the factory or a hospital, many problems are solved with principles of geometry. You've already learned about some of these principles in previous units of *Applied Mathematics*. You are going to learn more in this unit.

Figure 28-1 shows typical problems one can solve with geometry. The problems you'll meet in this unit will be more challenging.

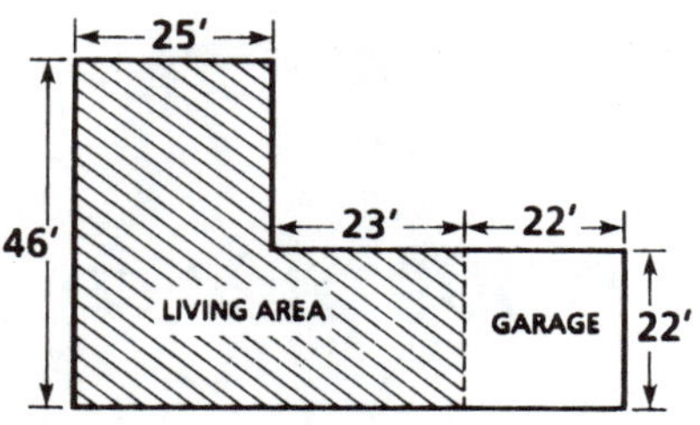

a. Using rectangles to measure floor area

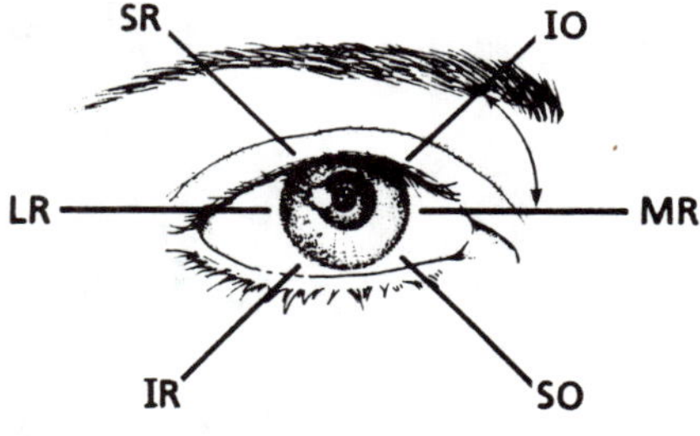

b. Using lines and angles to evaluate eye muscles

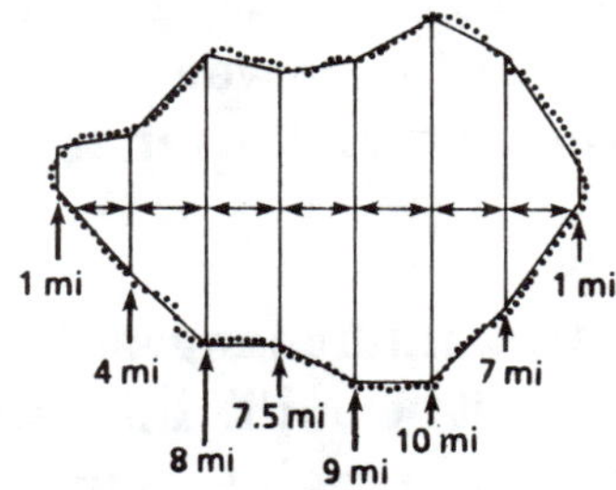

c. Using trapezoids to estimate the surface area of a lake

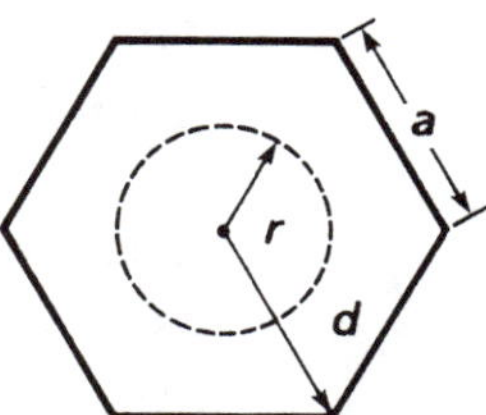

d. Using circles and hexagons to design a machine piece

Figure 28-1

Applying geometry to solve problems in the workplace

As you work through this unit, you will apply principles of geometry to solve real problems that occur in the workplace. You'll find yourself dealing with problems that involve:

- Rectangles, trapezoids, triangles, circles, sectors and segments of circles

- Tangent lines, perpendicular lines, and bisecting lines

- Special triangles (equilateral, right, 30°– 60°– 90°, and 45°– 45°–90°)

- Area, perimeter and circumference

- Pythagorean theorem

- Trigonometric functions of an angle (sin, cos, tan)

You'll apply geometry in two dimensions to solve real problems that occur in:

- Industrial technology

- Agriculture/agribusiness

- Business and marketing

- Home economics

- Health occupations

GEOMETRY IN INDUSTRIAL TECHNOLOGY

Industry manufactures products that range in design from simple items, like screws and screwdrivers, to complicated items, like engine parts and automobiles. Mass production in manufacturing keeps the cost of building these items down. Imagine, if you can, building an automobile by hand. The cost to us would be many hundreds of thousands of dollars—rather than the ten to twenty thousand we may spend now.

To achieve *mass production*, interchangeable parts of high quality and "exact" dimensions must be built. Interacting parts—like pistons and cylinders in an automobile engine—must fit together accurately. Parts that do not fit together lead to device failure and waste. To ensure an accurate fit of common parts, application of geometry, careful measurement, and quality construction must "work together."

Let's look at how geometry is involved in the design of some of the
simpler products turned out by industry.

Designing a common screwdriver blade

At one time or another, we all use a screwdriver. There are many
lengths available, with different handles and different size blade tips.
Geometry is involved in designing the blades.

Figure 28-2 shows a screwdriver with an expanded view of the tip.
Some of the blade dimensions are given—and some are missing.

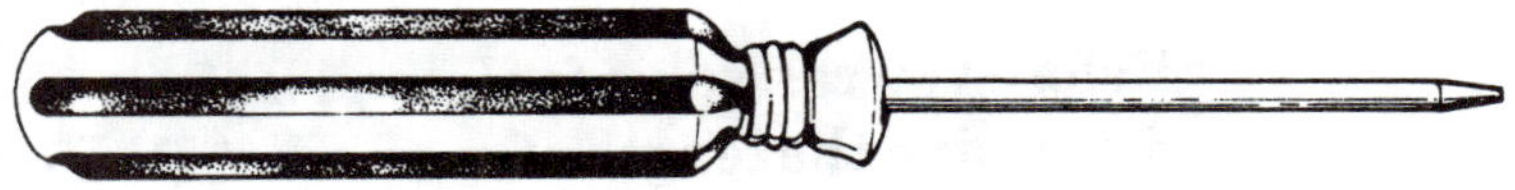

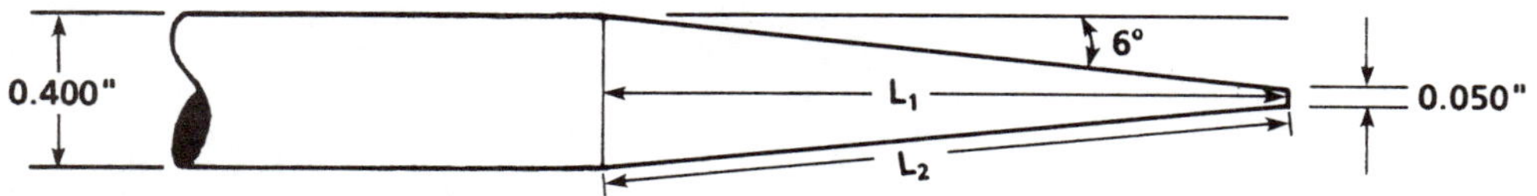

Figure 28-2
Common screwdriver with blade details

Example 1:
*Determining
dimensions for
a screwdriver
blade*

A designer makes a completed drawing of a screwdriver for a
machinist to follow. Part of the blade view shown in Figure 28-2 is
expanded and shown below. Before the machinist can make the blade,
dimensions L_1 and L_2 must be determined. How can they be found?

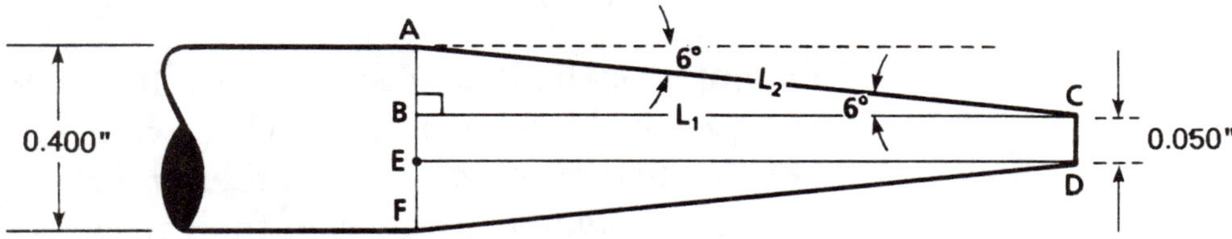

Right triangle **ABC** has been drawn. Note that it contains the
unknown length L_2 (side **AC**) as its hypotenuse and the unknown
length L_1 (side **BC**) as one of its legs. The angle **ACB** between L_1 and
L_2 is 6°.

You can find L_2 by using the *sine function*—since the sine of $\angle ACB$ relates side **AB** and hypotenuse L_2. Thus, since $\angle ACB = 6°$,

$$\sin 6° = \frac{\text{side opposite } 6°}{\text{hypotenuse}} = \frac{AB}{L_2}$$

In the same way you can find L_1 by using the *tangent function* which involves L_1, **AB**, and $\angle ACB$, as follows:

$$\tan 6° = \frac{\text{side opposite } 6°}{\text{side adjacent}} = \frac{AB}{L_1}$$

Before you can solve for L_1 and L_2 from these equations, you need to determine the length of segment **AB**. By studying the drawing, you can see that **AB** can be calculated, since **AB** = (**AF** − **BE**)/2, and both **AF** and **BE** are given. Thus

$$AB = \frac{AF - BE}{2}$$

$$AB = \frac{0.400 - 0.050}{2} = 0.175''$$

Knowing **AB**, and rearranging the equations above that involve L_1 and L_2, you can calculate as follows:

$$L_2 = AC = \frac{AB}{\sin 6°} = \frac{0.175}{0.10453} = 1.674'' \text{ (rounded to nearest thousandth)}$$

$$L_1 = BC = \frac{AB}{\tan 6°} = \frac{0.175}{0.10510} = 1.665'' \text{ (rounded to nearest thousandth)}$$

Note that if you had rounded *sin 6°* to 0.10 and *tan 6°* to 0.10, L_1 and L_2 would have had the same value. Thus, when dimensions are nearly equal, it is important to handle carefully the numbers you use in the calculations. Round off only your final answers.

 Calculate dimensions L_1 and L_2 to the nearest thousandth of an inch for a screwdriver whose blade width (**AF**) is 0.500 inch and blade tip (**CD**) is 0.080 inch. (Refer back to Example 1.)

CORD Applied Mathematics

The geometry of threads on a machine bolt

Figure 28-3 shows another everyday item—a *machine bolt*. Machine bolts and wood screws are common fasteners. Let's study the geometry necessary to make them.

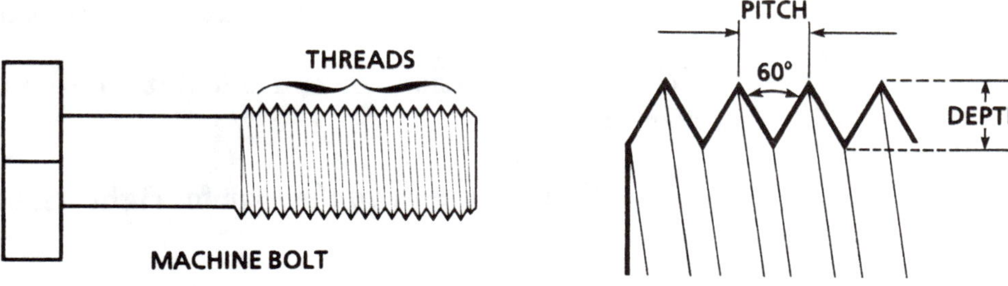

Figure 28-3
Machine bolt with thread details shown

As you can see in Figure 28-3, the sides of adjacent threads make angles of 60°. The *pitch* of the thread is the distance between adjacent crests. If you know the *number of threads per inch*, you can find the pitch by taking the reciprocal of that number. Thus

$$\text{Pitch} = \frac{1}{\text{No. of threads per inch}} = \frac{1}{\text{TPI}}$$

Example 2:
Calculating thread depth

Look at the enlarged view of the threads shown in Figure 28-4. Suppose you know that there are 20 threads per inch.

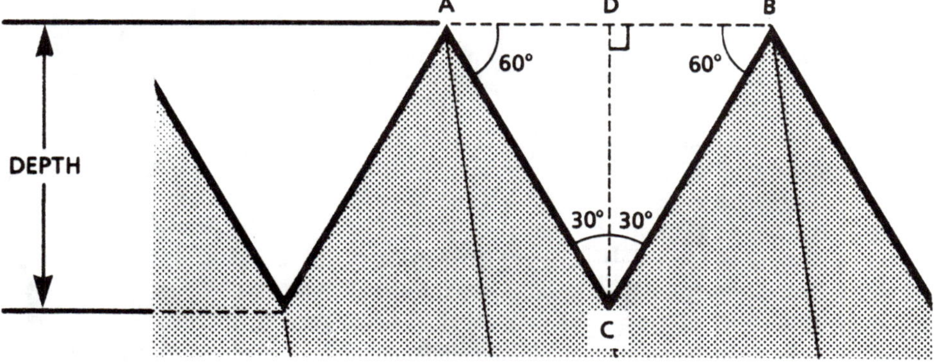

Figure 28-4
Enlarged view of thread geometry

From the drawing in Figure 28-4, you can observe the following:

- Since there are 20 threads per inch, the pitch ($^1/\text{TPI}$) is equal to $^1/_{20}$ inch or 0.050 inch. Thus **AB** = 0.050 inch.

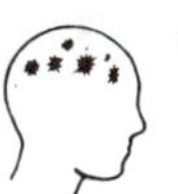

- The angle between the sides of the threads (∠**ACB**) is 60°. Therefore each half-angle (∠**ACD** or ∠**BCD**) is 30°.

- Since the angles **CAD** and **CBD** are also equal to 60°, the triangle **ACB** is *equilateral* (**AC** = **BC** = **AB** = 0.050").

- Triangles **ACD** and **BCD** are both right triangles.

With this information you can calculate the depth (**DC**) by either of two methods.

a. Use the *Pythagorean theorem* for right triangle **BCD** to find **DC**:

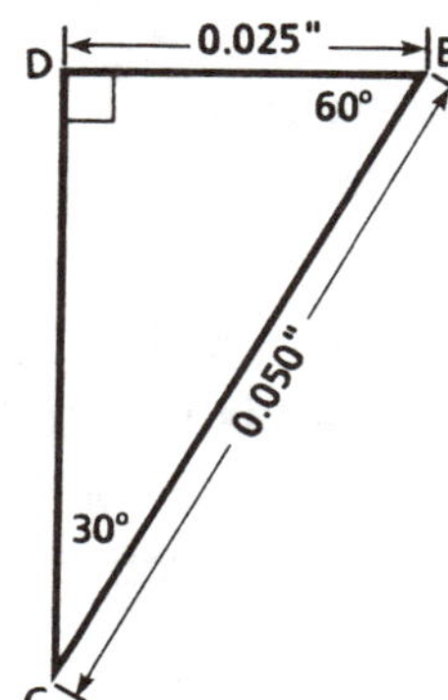

$$(\textbf{HYP})^2 = (\textbf{LEG})^2 + (\textbf{LEG})^2$$

$$(\textbf{CB})^2 = (\textbf{BD})^2 + (\textbf{DC})^2; \quad \textbf{BD} = {}^1\!/_2\,\textbf{AB} = 0.025"$$

$$(0.050)^2 = (0.025)^2 + (\textbf{DC})^2$$

$$\textbf{DC} = \sqrt{(0.050)^2 - (0.025)^2}$$

$$\textbf{DC} = 0.043 \text{ inch}$$

Thus, the thread depth is found to be 0.043 inch (rounded to the nearest thousandth of an inch).

b. Use *trigonometric functions* for right triangle **BCD** to find **DC**:

$$\tan 30° = \frac{0.025"}{\textbf{DC}} \qquad\qquad \cos 30° = \frac{\textbf{DC}}{0.050"}$$

$$\textbf{DC} = \frac{0.025"}{\tan 30} \qquad\qquad \textbf{DC} = (0.050")\,(\cos 30°)$$

$$\textbf{DC} = \frac{0.025"}{0.57735} \qquad\qquad \textbf{DC} = (0.050")\,(0.86603)$$

$$\textbf{DC} = 0.043 \text{ inch} \qquad\qquad \textbf{DC} = 0.043 \text{ inch}$$

Again, the thread depth is found to be 0.043 inch.

An actual American Standard thread has a *flat* on the top and bottom, as shown in the drawing in Figure 28-5. The flat is $1/8$ of the pitch size. The angle between the thread sides is 60°. There are 20 threads per inch, so the pitch is $1/\text{TPI} = 1/20 = 0.050$ inch. The thread depth before the flats are made is 0.0433 inch, as calculated earlier in Example 2.

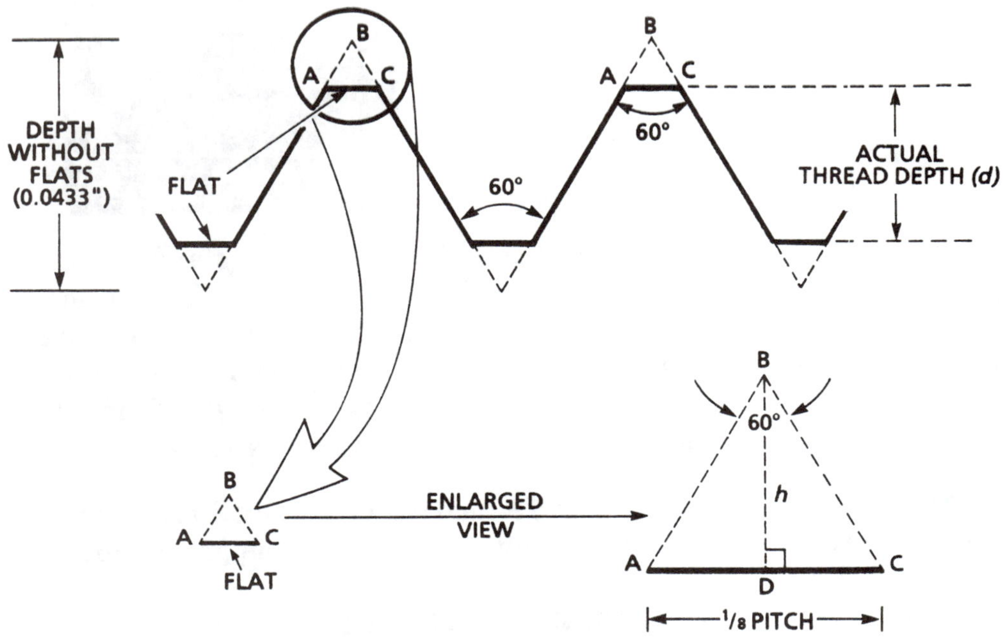

Figure 28-5
Geometry for a 20 TPI American Standard thread

Study the diagrams above. Determine the actual thread depth d. To help you solve this activity, consider the following questions:

- What is the relationship between 1) the 0.0433-inch *depth without flats*, 2) the segment h in triangle **ABC**, and 3) the actual thread depth d?

- How can you use the properties of right triangle **CBD** or **ABD**—in the enlarged triangle **ABC**—to determine the length of h?

- If line **BD** in triangle **ABC** is drawn perpendicular to the base **AC**, what is the relationship between segment **DC** and base **AC**?

Carry out your calculations to 5 or 6 decimal places, **then round** your final answer to the nearest ten-thousandth of an inch.

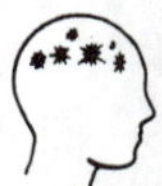

All geometry in the industrial workplace does not necessarily involve mechanical items. The drawing below is something most of us see every day—the television.

The television industry has set a standard for the size of picture tubes. The ratio of *height-to-width* is 3:4. This ratio is called the *aspect ratio*. It means, for example, that a television picture tube 30 inches high will be 40 inches wide. Television *picture-tube size* is given in accordance with the *diagonal length* of the tube face. Thus, a nineteen-inch television set has a face with a 19-inch diagonal.

Example 3:
**Dimensions
of a television
screen**

You own a 19-inch television set. Based on the aspect ratio of 3:4 for height-to-width, what are the height (h) and width (w) of your set?

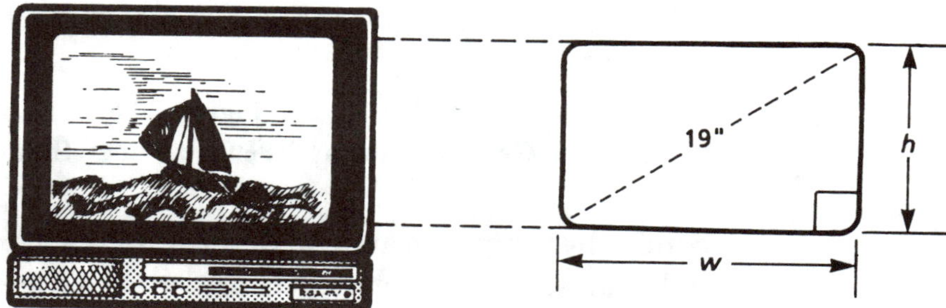

Use the right triangle formed by the 19" diagonal, the width (w) and the height (h). Applying the Pythagorean theorem,

$$w^2 + h^2 = (19)^2$$

Since $\dfrac{h}{w} = \dfrac{3}{4}$, it is true that $h = \dfrac{3}{4}\,w$, or $h = 0.75\,w$.

Substituting $0.75\,w$ for h, the Pythagorean theorem gives us

$$w^2 + (0.75\,w)^2 = (19)^2$$

Simplifying,

$$w^2 + 0.5625\, w^2 = 361$$

From this you can solve for w to get

$$w = \sqrt{361/1.5625} = 15.2 \text{ in. (rounded to tenth of an inch)}$$

And since $h = {}^{3}/_{4}\, w$

$$h = 0.75\,(15.2) = 11.4$$

Thus the dimensions ($h \times w$) for a 19-inch television set screen are about 11.5" $\times$ 15".

Study Activity:

Your company manufactures television tubes for the industry. You receive an order for television tubes that specifies a width of 24 inches. What are the height and diagonal measurements for the tube?

Geometry in the tool and die industry

Most every sheet-metal part manufactured today has some type of hole in it. This hole is stamped from sheet metal by a tool that involves a punch and a die. Any type of equipment used to manufacture parts is called a *tool*. (These are not the hand tools that you are probably familiar with.) In manufacturing, the words *tool* and *tooling* describe a wide variety of dies, punches, jigs, fixtures, molds, special machines, and precision parts that are used to make almost everything we use in everyday life.

Figure 28-6 shows pictures of several common punches used to stamp holes in sheet metal. The cross-sectional shapes at the ends of the punches shown include a circle, a square and a triangle.

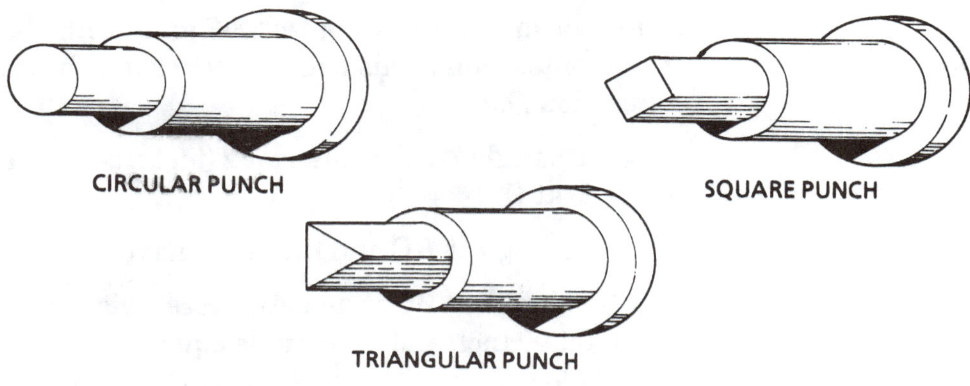

Figure 28-6
Punches for metal stamping

In metal stamping, the punch pushes its way through the metal workpiece. The workpiece itself lies on top of a die—a support sheet with a hole beneath the metal exactly where the punch is to push through. To make a clean cut through the metal, the "hole" in the die must match accurately the cross-sectional shape of the punch.

Example 4:
Designing a triangular punch

Suppose a *triangular punch* is to be ground from a round metal bar. The triangular shape you want for the end of the punch is an *equilateral triangle*, with each side equal to 0.500 inch. See drawing in Figure 28-7. You want to minimize waste and work time as you grind the bar down to the triangular cross section. What is the diameter of the circular bar stock you should select?

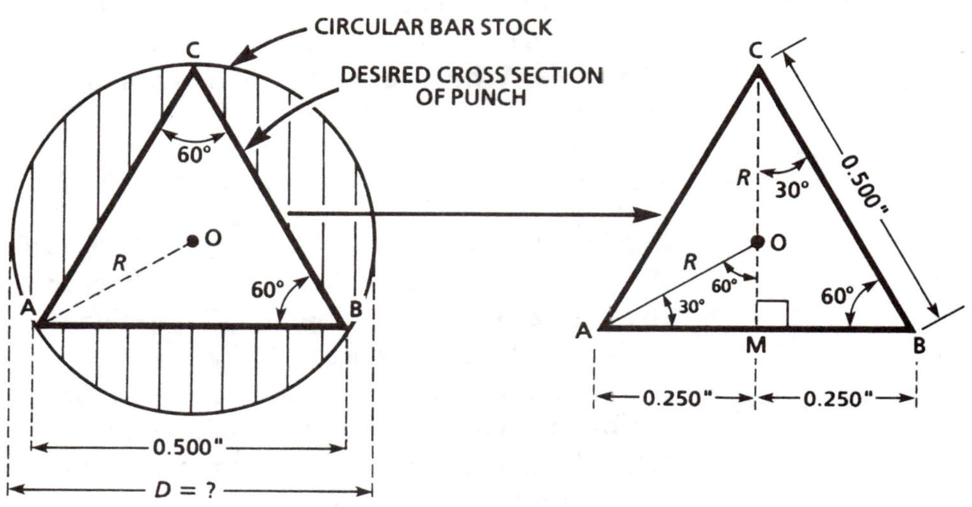

Figure 28-7
Geometry involved in shaping a triangular punch

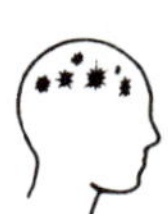

Study the drawings in Figure 28-7 and think about the following statements. Doing this will help you solve for the required dimension *D*.

- Triangle **ABC** is *inscribed* (just fits) in the circle whose center is at **O**.

- Triangle **ABC** and the circle have the same center **O**.

- Triangle **ABC** is an equilateral triangle with each side equal to 0.500 inch and each angle equal to 60°.

- A line segment from point **O** to any of the three vertices of the triangle (**A**, **B**, or **C**) bisects the 60°angle and forms two 30° angles.

CORD Applied **Mathematics**

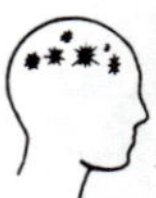

- The line segment (**OA**, **OB** or **OC**) from point **O** to any of the vertices of the triangle is equal to the radius R of the circle.

- The height (line segment **CM**) of $\triangle$**ABC** is perpendicular to the base (**AB**) of the triangle and divides line segment **AB** into two equal parts.

- The height **CM** is part of right triangle **BCM**, with right angle at point **M**. Thus, from the Pythagorean theorem, $(\mathbf{CM})^2 + (\mathbf{BM})^2 = (\mathbf{BC})^2$.

- Line segment **MO** equals $^1/_2$ of the radius R or $^1/_3$ of the altitude (height) **CM**.

- Triangle **AMO** is a right triangle that involves line segments **AM**, **MO** and R. One side (**AM**) and all angles are known.

There is more than one way to solve for R, the unknown. If you have thought about the steps above, you can solve for R by either of the two methods that follow.

Method 1: Use the 30° – 60° – 90° right triangle **AMO** to set up the Pythagorean relation $R^2 = (\mathbf{AM})^2 + (\mathbf{MO})^2$. The line segment **AM** is equal to 0.250 inch. The line segment **MO** is not known, but it is equal to one-third of the altitude or height (line segment **CM**). Line segment **CM** can be found by using the *Pythagorean theorem* or appropriate *trigonometric function* (tangent or cosine) with respect to the 30° – 60° – 90° right triangle **BCM**. For $\triangle$**BCM**, the hypotenuse (**CB**) and base leg (**MB**) are both known. Thus, since **CM** can be calculated, and $\mathbf{MO} = {}^1/_3(\mathbf{CM})$, $R = \sqrt{(\mathbf{AM})^2 + (\mathbf{MO})^2}$.

Method 2: Use the 30° – 60° – 90° right triangle **AMO**. Since all angles and one side (leg **AM**) are known, the hypotenuse R can be found from either of the two relations that follow:

$$\sin 60° = \frac{\text{side opp}}{\text{hyp}} = \frac{0.250}{R} \qquad \cos 30° = \frac{\text{side adj}}{\text{hyp}} = \frac{0.250}{R}$$

Method 2 is the shorter, more direct way to solve for R. But note clearly that whichever method you use, *an important part of your solution is to be able to draw and label the triangles shown in Figure 28-7.*

Using either method, you should find a value for R equal to 0.2887 inch (rounded) and hence a diameter $D = 2R$ of 0.5774 inch (rounded). Since it's doubtful if bar stock of this dimension is available, you would probably have to begin with $^5/_8$-inch-diameter (0.625 inch) bar stock for the grinding process.

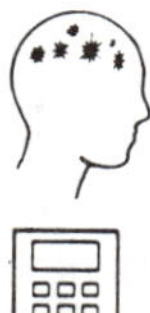

Use what you learned from Example 4 to determine the size of a square that fits exactly inside (inscribed within) a circle of 1-inch diameter, as shown here.

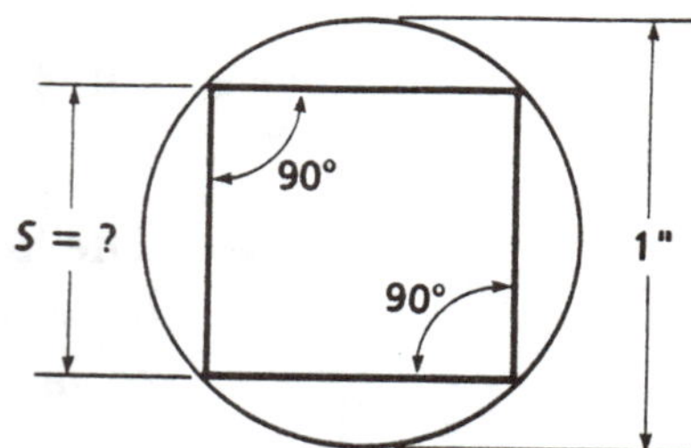

Can you draw a diameter of the circle to form 45°– 45°– 90° right triangles within the square? Can you use the properties of one of the 45°– 45°– 90° right triangles to determine the sides of the square?

GEOMETRY IN AGRICULTURE

Often in agriculture one needs to predict crop yields, apply pesticides and fertilizers, or purchase additional land. To do this, one must be able to calculate the areas of differently shaped plots of land.

It's a fairly easy task to calculate the area of a field that is a rectangle. For instance, a rectangular field 500 feet by 1000 feet has an area of 500,000 square feet. Area measurements of land in the U.S. are generally expressed in acres. An acre is equal to 43,560 square feet. Therefore, to calculate the area of a 500-foot by 1000-foot field in acres, you would multiply 500 times 1000 and divide by 43,560. The result to the nearest half acre is 11.5 acres.

This is a relatively simple calculation, involving the formula $(A = L \times W)$ for the area of a rectangle. In many cases, fields are not rectangular, so area calculations can be more complicated. Example 5 shows how to determine the area for a nonrectangular field.

Calculating areas for nonrectangular shapes

Example 5:
Calculating area for a trapezoidal field

A surveyor took sightings for a plot of land shaped like that shown below. The sketch contains the results of the surveyor's measurements.

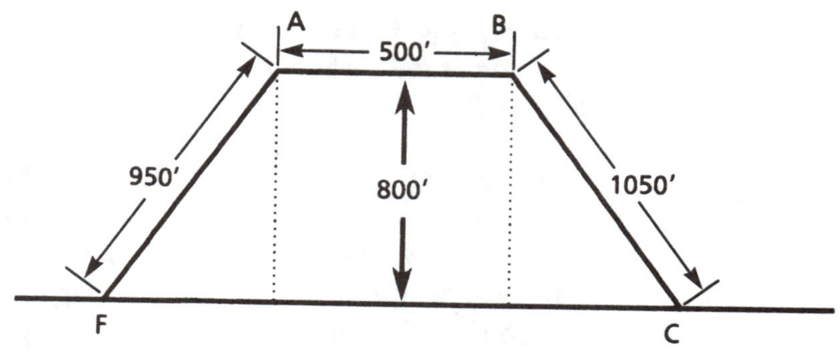

CORD Applied Mathematics

From an analysis of the drawing you can see that:

- The plot of land has four sides (**AB**, **FC**, **AF** and **BC**), only two of which are parallel (**AB** and **FC**). Thus it is a trapezoid.

- The area of a trapezoid is given by the formula

$$A = \frac{1}{2} \times (b_1 + b_2) \times h$$

where b_1 and b_2 are the lengths of the two parallel sides and h is the perpendicular distance between them.

- For the given sketch, the area is

$$A = \frac{1}{2} \times (\mathbf{AB} + \mathbf{FC}) \times h$$

where **AB** = 500' and h = 800'. The length of base **FC** is not given, but it can be found from other information that is given, as follows:

- The trapezoid can be divided into two triangles and a rectangle by drawing perpendicular lines **AE** and **BD**.

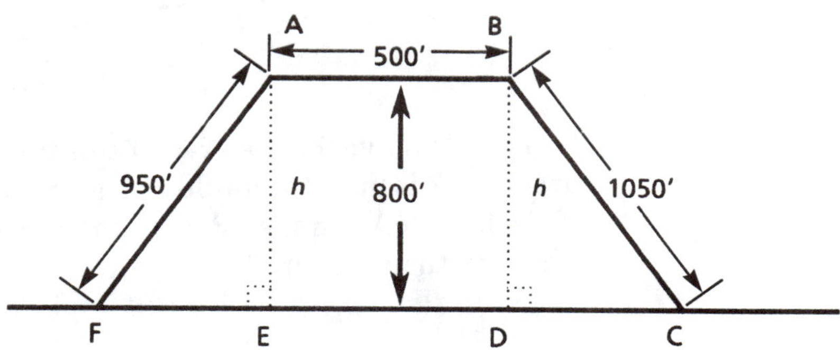

- Line segments **AE** and **BD** (labeled as h) are each equal to 800'.

- Line segment **EF** in Δ**AEF** and line segment **DC** in Δ**BDC** can be found by using the *Pythagorean theorem*.

- If **FE**, **ED** and **DC** are known, **FC** can be calculated. If **FC** is known, the area of the trapezoid can be calculated from the formula $A = \frac{1}{2}(\mathbf{AB} + \mathbf{FC}) \times h$.

Carry out the necessary steps to determine the area of the trapezoidal field **ABCF**, in Example 5, and give your result in acres rounded to the nearest tenth of an acre. Did you get $A = 20.1$ acres (about 876,960 ft²)?

Notice again the importance of drawing *auxiliary lines*—such as **AE** and **BD**—to help you use the given information to solve the problem.

Calculating areas that involve triangles and circles—or parts of circles—can be even more challenging. Look at a sketch of the piece of land shown below. Points **B** and **D** are points of *tangency* between the straight line and the circular arc at the end.

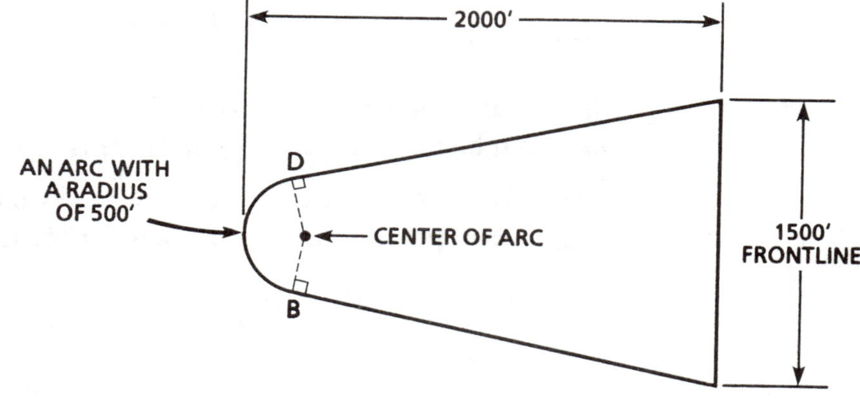

In order to solve for the area of this nonregular shape, you should first break it up into smaller regular shapes (triangles, circles, sectors of circles, etc.) and then determine the area of each of these shapes. One way to divide up the large area is shown in Figure 28-8.

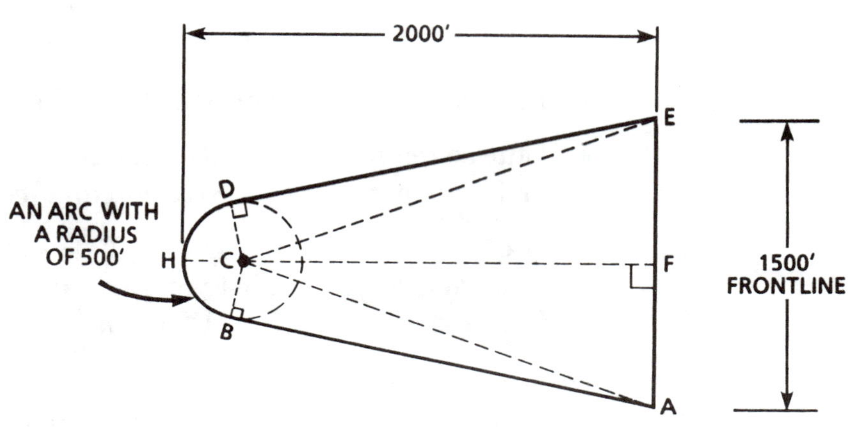

Figure 28-8
Subdividing a shape into standard pieces

Note carefully that the nonregular shape (**ABDE**) has been divided into the following parts:

- Two equal right triangles, **ACF** and **ECF**

- Two equal right triangles, **CDE** and **CBA**

- A sector of a circle, **BCD**, made up of two equal sectors, **BCH** and **DCH**

The problem now is simpler. From the data given, you can calculate the areas of the separate pieces. In fact, to make it still easier, notice that the pieces above **symmetry** line **FH** have the same areas as those below **FH**. So you can calculate the total area of the pieces located above the symmetry line **FH** and double to get the overall area.

Before you do this, study Figure 28-8 and make sure you understand each of the statements that follow. Read them carefully and refer to Figure 28-8 to help you understand their meaning.

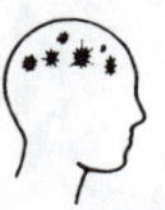

- Point **C** is the center of a circle of which the arc **BHD** is a piece.

- Line segment **ED** meets one end of the arc at point **D**. At that point, **ED** is *tangent* to the circle with center at **C**. The radius **CD** is perpendicular to line segment **ED** at the *point of tangency*. The sketch drawn below shows the relationship that exists ALWAYS between a radius and tangent line at the point of tangency. Thus, ∠**CDE** in Figure 28-8 is a right angle and Δ**CDE** is a right triangle.

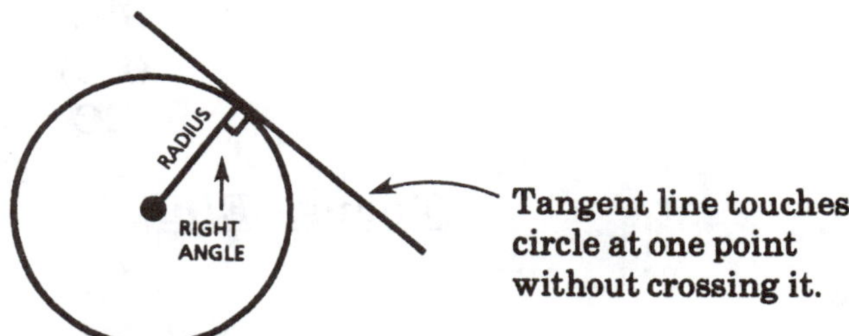

- In the same way, line segment **AB** is tangent to the circle at **B**, and radius **CB** is perpendicular to **AB** at point **B**. Thus ∠**CBA** is a right angle and Δ**ABC** is a right triangle. Triangles **CDE** and **ABC** are *congruent* triangles and have equal areas.

- Line segment **FH** is a line of symmetry that divides the overall shape into two equal parts. This line segment is drawn perpendicular to line segment **AE** at **F** and divides **AE** into two equal parts. Triangles **ACF** and **ECF** are *congruent* right triangles, equal to each other in area.

- Sector **BCD** represents a fraction of the area of the entire circle. The ratio of the area of sector **BCD** to the area of the circle is equal to the ratio of $\angle$**BCD** to 360°. So,

$$\text{Area of sector} = \frac{\angle \mathbf{BCD}}{360°} \times \text{Area of circle}$$

- Angle **BCD** is equal to $[360° - 2\,(\angle \mathbf{ECD} + \angle \mathbf{FCE})]$.

Now you are ready to start the calculations for the area of the pieces. Refer to Figure 28-8 as needed. Do these calculations in order and be patient. This is **NOT** a short problem!

Triangle ECF

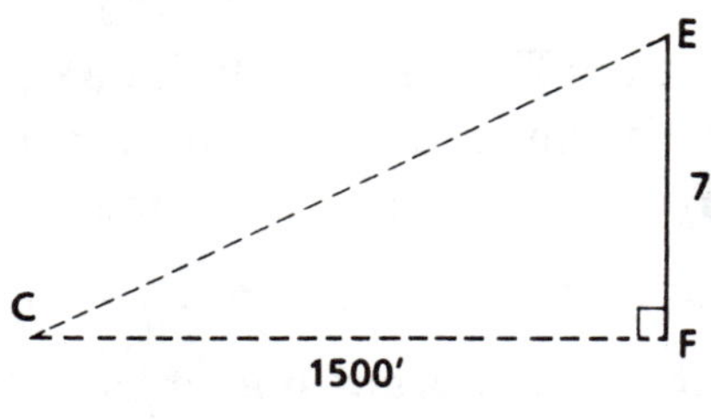

$$CF = HF - HC = 2000' - 500' = 1500'$$

$$EF = \frac{EA}{2} = \frac{1500}{2} = 750'$$

$$Area = {}^{1}\!/_{2}(b \times h) = {}^{1}\!/_{2}(\mathbf{EF} \times \mathbf{CF})$$

$$A = {}^{1}\!/_{2}(750)(1500) = \underline{562{,}500\ \text{ft}^2}$$

Line segment EC

You will need the length of line segment **EC** in order to find the area of $\triangle$**EDC**. Thus, from right triangle **ECF** above, you can write:

$$(\mathbf{EC})^2 = (\mathbf{CF})^2 + (\mathbf{EF})^2$$
$$(\mathbf{EC})^2 = (1500)^2 + (750)^2$$
$$\mathbf{EC} = \sqrt{(1500)^2 + (750)^2} = 1677\ \text{ft}$$

Triangle EDC

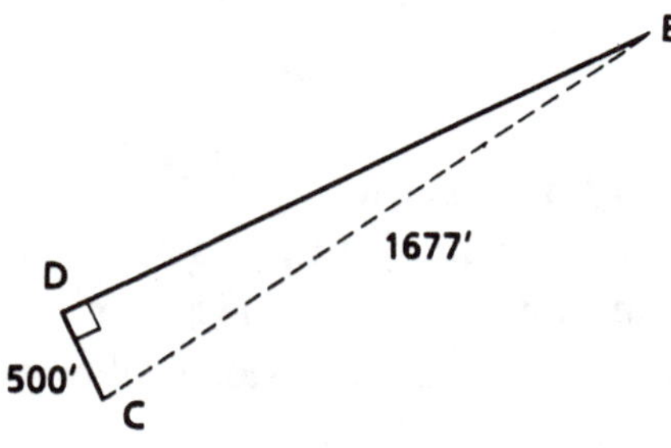

$$Area = {}^{1}\!/_{2}(b \times h)$$

$$A = {}^{1}\!/_{2}(\mathbf{CD} \times \mathbf{ED})$$

Obtain **ED** from the Pythagorean theorem

$$(\mathbf{ED})^2 = (\mathbf{CE})^2 - (\mathbf{CD})^2$$

$$\mathbf{ED} = \sqrt{(1677)^2 - (500)^2} = 1600.8\ \text{ft}$$

So, $A = {}^{1}\!/_{2}(500)(1600.8) = \underline{400{,}200\ \text{ft}^2}$

Angles ECD and FCE

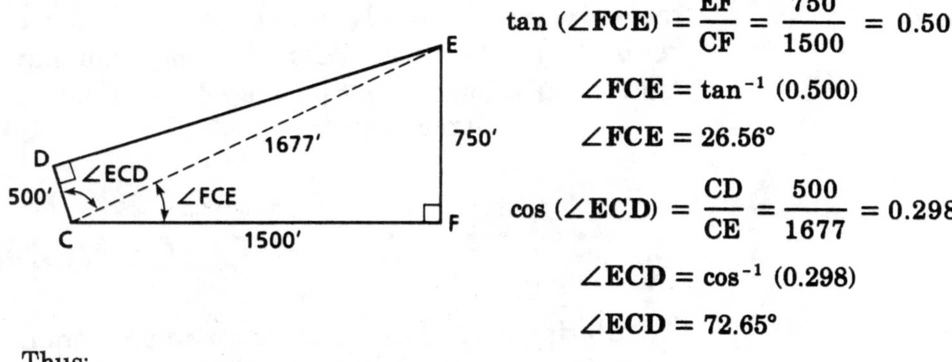

$$\tan(\angle FCE) = \frac{EF}{CF} = \frac{750}{1500} = 0.500$$

$$\angle FCE = \tan^{-1}(0.500)$$

$$\angle FCE = 26.56°$$

$$\cos(\angle ECD) = \frac{CD}{CE} = \frac{500}{1677} = 0.298$$

$$\angle ECD = \cos^{-1}(0.298)$$

$$\angle ECD = 72.65°$$

Thus:

$$\angle FCD = \angle FCE + \angle ECD$$
$$\angle FCD = 26.56° + 72.65°$$
$$\angle FCD = \underline{99.21°}$$

$$\angle BCD = 360° - 2(\angle FCD)$$
$$\angle BCD = 360° - 2(99.21°)$$
$$\angle BCD = 161.6° \text{ (rounded)}$$

Area of sector BCD

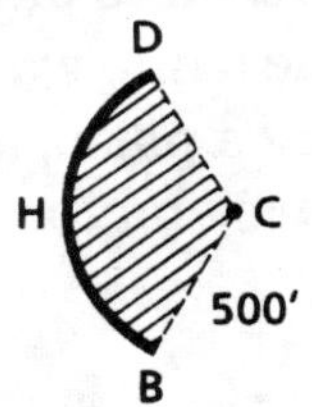

$$\text{Area of sector} = \frac{\angle BCD}{360°} \times \text{Area of circle}$$

$$\text{Area of circle} = \pi r^2 = \pi(500^2) = 785{,}398 \text{ ft}^2$$

So,

$$\text{Area of sector} = \frac{161.6°}{360°} \times 785{,}398 \text{ ft}^2 = \underline{352{,}556 \text{ ft}^2}$$

Total Area

$$A_{tot} = 2(\text{Area of } \Delta ECF) + 2(\text{Area of } \Delta EDC) + (\text{Area of sector BCD})$$

$$A_{tot} = 2(562{,}500) + 2(400{,}200) + (352{,}556)$$

$$A_{tot} = 1{,}125{,}000 + 800{,}400 + 352{,}556 = \underline{2{,}277{,}956 \text{ ft}^2}$$

$$A_{tot} = \frac{2{,}277{,}956 \text{ ft}^2}{43{,}560 \text{ ft}^2/\text{acre}} = 52.3 \text{ acres (rounded)}$$

As you can see, area calculations can become involved. But if you break a complicated figure into smaller regular shapes that you recognize, and then solve for the areas of them **one at a time**, you can get to the final answer. Along the way, you may have to make several "side calculations" to get missing dimensions.

GEOMETRY IN HOME ECONOMICS

Calculating areas is as common in home economics as it is on the farm. One definition of the word "economics" means using materials without waste. To make the best use of materials available, or to order materials correctly, calculating areas is a necessary task.

It does not matter whether you are planning to put tile, carpet, or wood on a floor, you will need to calculate the square footage of the area to order materials. If you intend to paint, wallpaper, or panel an area, calculations are made in the same way. If you are fertilizing the lawn, purchasing sod, or planning a patio, area calculations are necessary to calculate usable space and to order the right amount of materials.

Many times when one is involved in remodeling a home, the cost of the materials determines what one uses. Most of us want the best materials, but we are usually limited by a budget. When we calculate the square footage accurately, it is much easier to determine the type of material we can afford.

Let's calculate a sample floor area, so that we can order the right amount of carpet. For a simple rectangular floor the problem is easy. Suppose for example, the floor is 12 feet × 16 feet. Then

$$A = 12 \times 16 = 192 \text{ ft}^2$$

Since 1 square yard is equal to 9 square feet, we can get the area in square yards simply by dividing by 9.

$$A = \frac{192 \text{ ft}^2}{9 \text{ ft}^2/\text{yd}^2} = 21.3 \text{ sq yd}$$

Usually you round up to the nearest yard. So to carpet the 12' × 16' floor, you'd need about 22 square yards.

Rooms, walls, floors—or plots of ground—are seldom perfect rectangles. As previously discussed, the more expensive the material
used, the more you need accurate calculations. Sometimes, areas may
be estimated. Availability of materials is usually what determines
the method used.

Using graphical methods to estimate an irregularly shaped area

For irregularly shaped surfaces that cannot be broken down into
regularly shaped figures, a graphical method, using scaled drawings,
may work.

Example 6:
*Estimating
an area
by a graphical
method*

Suppose you need to find the approximate area of an oddly shaped
floor, such as the one shown below. After studying the figure you
will probably decide you *cannot break it down* into a set of *rectangles,
triangles, circles, sectors...* as you did before. How then can you find
the approximate area?

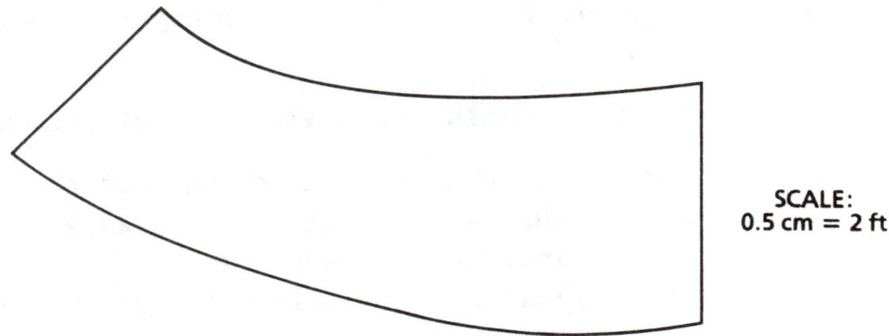

The figure shown above has been drawn to scale. You are told that
the scale is 0.5 cm = 2 feet. Approximately how many square feet of
area are there?

First, lay out a grid with 0.5-cm squares covering the area, as shown
below.

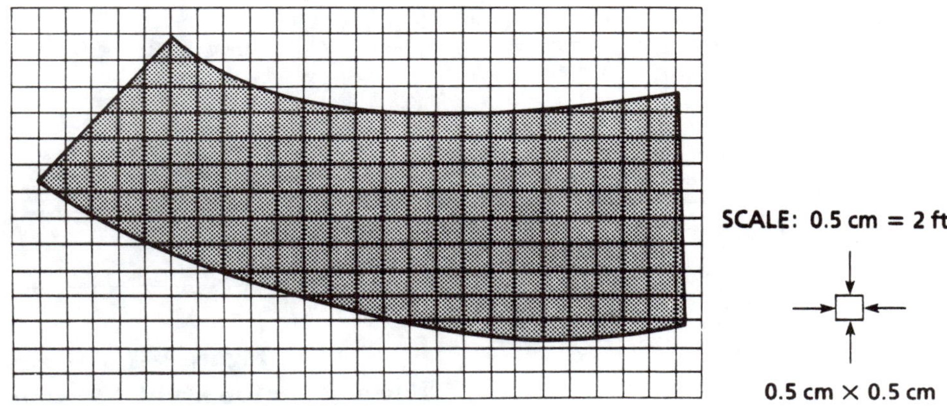

To get an estimate for the area, you count squares. First count all the
full squares (completely shaded). Then count the number of **incomplete** squares (partially shaded). Take $1/2$ of the partially shaded
squares to get an estimate for the equivalent number of complete
squares. Then add this to the number of complete squares counted to
get the number of total squares.

If you do this for the shape we're working on here, you get:

> 157 complete squares
>
> $1/2 \times$ (52 partially shaded squares) $\approx$ 26 "complete squares"

Thus, there are close to $157 + 26 = 183$ "total" squares. Since each
0.5-cm square represents a $2' \times 2'$ area, each square is 4 ft². The area
of the odd shape is then

$$A = 183 \text{ squares} \times 4 \text{ ft}^2/\text{square}$$
$$A = 732 \text{ ft}^2 \text{ (approximately)}$$

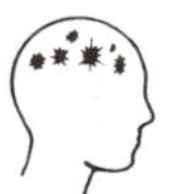

Admittedly, the area you have determined is an *estimate*. Using this
same graphical method, how could you improve the accuracy of your
estimate?

Finding areas that involve segments of circles

You are part of a home-remodeling team that has been asked to lay
new oak flooring for the room shown below. Except for the bay-
window area, the room is a rectangle, $12' \times 28'$. Rectangular areas
are easy to calculate. But how about the bay-window area?

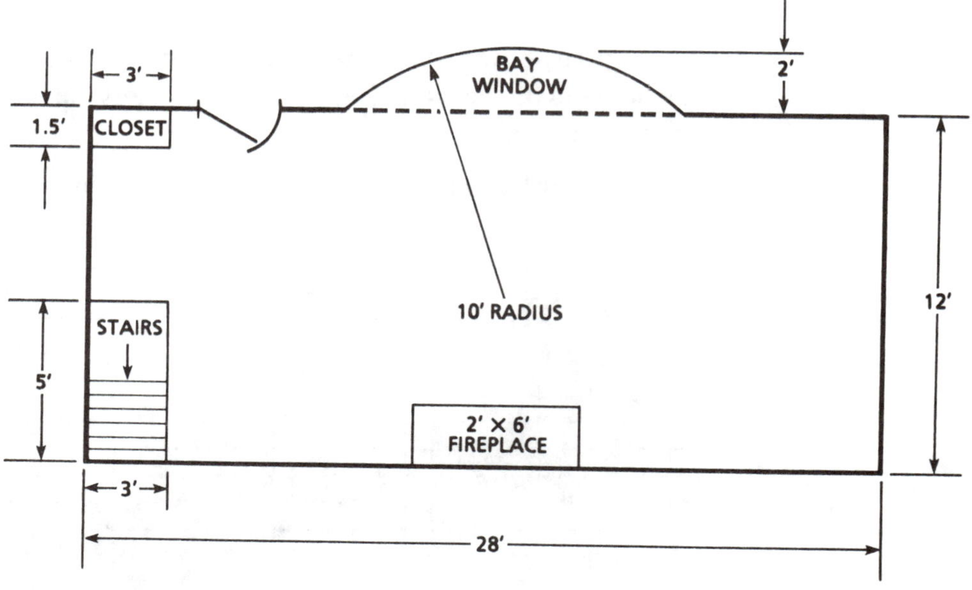

Figure 28-9
Rectangular room with bay window

CORD Applied Mathematics

According to Figure 28-9, the outer edge of the bay window is part of a circle of 10-ft radius. The "depth" of the bay window is 2 ft.

The bay-window area (shaded) is redrawn below—as part of the circle it belongs to—with appropriate dimensions.

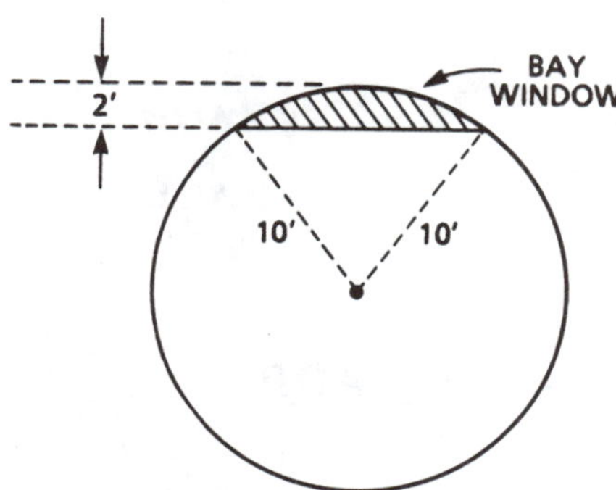

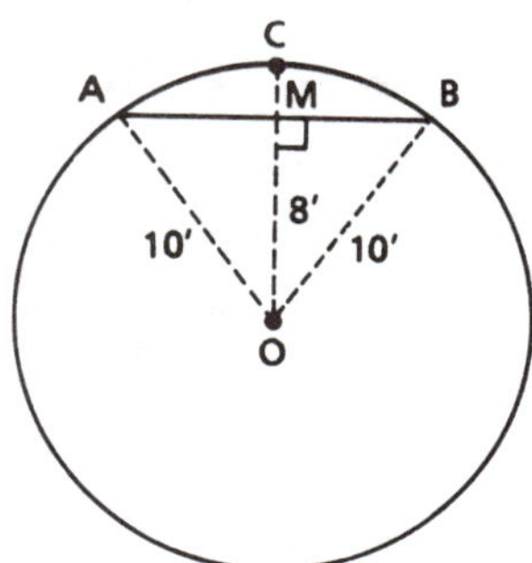

In the language of mathematics, the bay-window area (shaded in the drawing on the left) is called the **segment of a circle**. The *line segment* **AB** is called a **chord**, and *line segments* **AO**, **CO**, and **BO** are each equal to the radius of the circle.

To find the area of the bay window, you first find the area of the circle sector (**AOB**), next find the area of the triangle **ABO**, and then subtract the area of the triangle from the area of the sector. The result is the area of the segment—the bay window.

To find the area of the sector and triangle, you will need to know the dimension **AB**. To find **AB**, look at right triangle **BOM**. It has a hypotenuse of 10' and a leg of 8'. Using the Pythagorean theorem, you find that the leg **BM** is 6 ft long. Since **AM = BM**, **AM** is also 6 feet in length. Now you have all the dimensions you need to solve the problem.

Before you begin the solution, study the drawing above and make sure you understand the following statements.

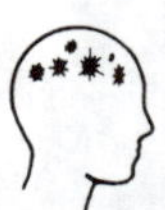

- The radial line segment **CO** bisects ∠**AOB**, forming two equal angles, ∠**AOC** and ∠**BOC**.

- The radial line segment **CO** is perpendicular to the chord **AB** at **M**, and divides **AB** into two equal parts, **AM** and **BM**.

- The area of the sector (**AOB**) is equal to the fraction (∠**AOB**/360°) multiplied by the area of the 10-ft circle.

Determine $\angle MOB$ and $\angle AOB$

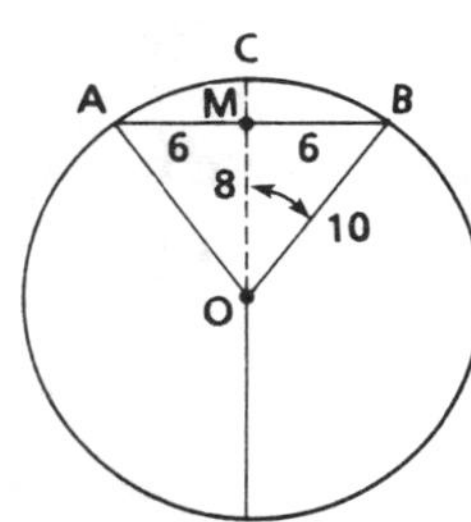

$$\sin\ (\angle MOB) = \frac{MB}{OB} = \frac{6}{10} = 0.6$$

$$\angle MOB = \sin^{-1}(0.6)$$

$$\angle MOB = 36.87°\ \text{(rounded)}$$

$$\angle AOB = 2 \times \angle MOB$$

$$\angle AOB = 2 \times 36.87°$$

$$\angle AOB = 73.7°\ \text{(rounded)}$$

Determine area of sector AOB

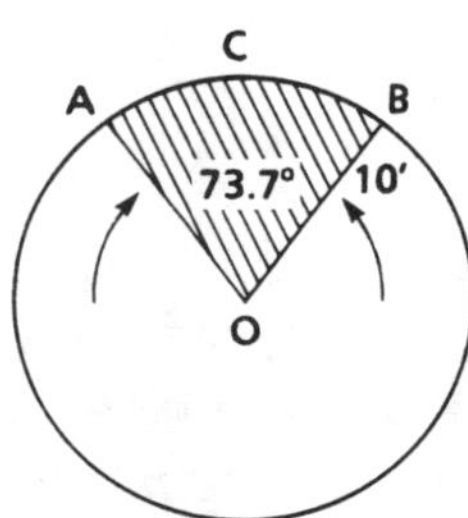

$$\text{Area of sector AOB} = \frac{\angle AOB}{360°} \times \text{Area of circle}$$

$$A_{\text{sector}} = \frac{73.7°}{360°}(\pi)(10)^2$$

$$A_{\text{sector}} = (0.2047)(\pi)(100)$$

$$A_{\text{sector}} = \underline{64.31\ \text{ft}^2}$$

Determine area of $\triangle ABO$

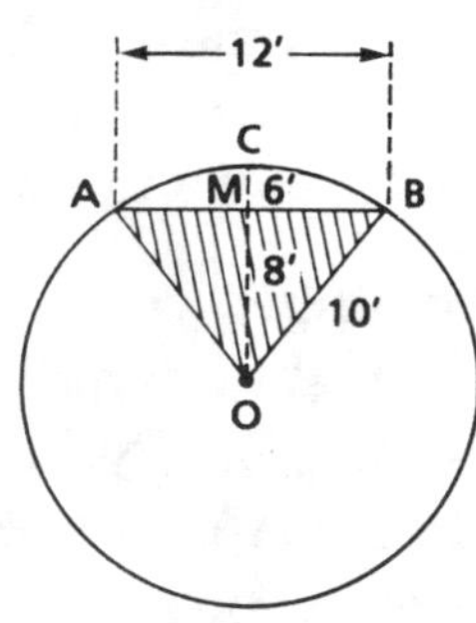

$$\text{Area of } \triangle ABO = \tfrac{1}{2} \times AB \times OM$$

Here $AB = 12'$ and $OM = 8'$, so . . .

$$\text{Area of } \triangle ABO = \tfrac{1}{2} \times 12 \times 8 = \underline{48.00\ \text{ft}^2}$$

Determine area of segment ABC

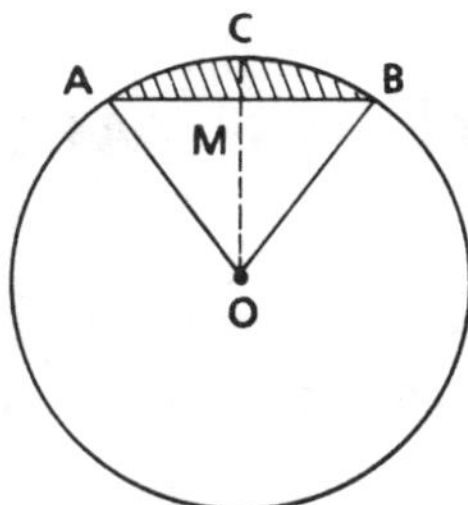

$$\text{Area (ABC)} = \text{Area of sector \textbf{AOB}} - \text{Area of } \triangle \textbf{ABO}$$

$$\text{Area (ABC)} = 64.31\ \text{ft}^2 - 48.00\ \text{ft}^2$$

$$\text{Area (ABC)} = \underline{16.31\ \text{ft}^2}$$

Thus, the area of the bay window is 16.31 ft²

Note: There is a formula that gives the area of a segment of a circle directly. If r is the radius of the circle and θ is the sector angle spanned by the segment, then the area of the segment is

$$A = \frac{\pi r^2 \theta}{360°} - \frac{r^2 \sin \theta}{2}$$

For our problem,

$$r = \mathbf{AO} = 10'$$

$$\theta = \angle \mathbf{AOB} = 73.7°$$

Thus, substituting in the values for r and θ, you can write

$$A = \frac{\pi (10)^2 (73.7°)}{360°} - \frac{(10)^2 \sin (73.7°)}{2}$$

Carry out this calculation to see if you get the same result as above.

Refer back to Figure 28-9. Now that you have found the area of the bay window, it should not be too difficult to calculate the total square feet of floor area to be covered. Assuming that you will **not cover** the space for fireplace, closet, and the area taken out by the stairs, what is the total area to be covered?

GEOMETRY IN BUSINESS AND MARKETING

There are many applications of geometry in business and marketing. Computing square footage of marketplaces and selling areas is necessary to obtain the best use of available space. Using charts and geometry to forecast business trends is a daily activity. One of the largest uses of geometry in business and marketing is in the packaging industry.

Packaging of products must be done as inexpensively as possible. However, the packages themselves must be strong, durable, and pleasing to the eye. As you work through this unit, spend some time looking at packages that you use daily. Candy wrappers, cereal boxes, milk containers, and pop cans are but a few of the items made possible by geometry in the workplace.

You are part of a design team that produces labels for soup cans. The can is 3 5/8" high and 2 5/8" in diameter. You plan to allow a 1/4-inch overlap to glue the ends of the label together. What length of paper is required per can label?

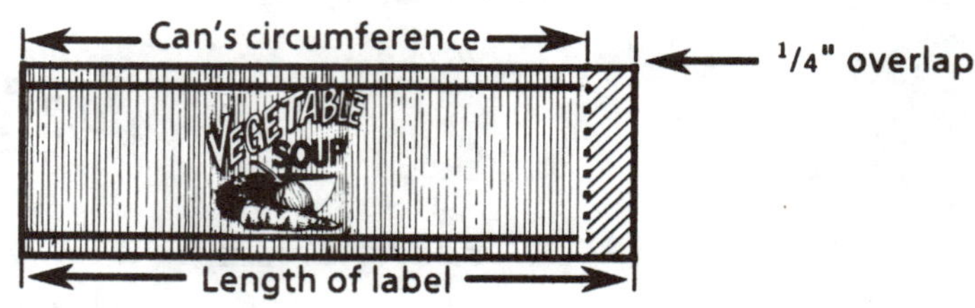

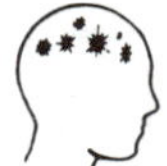

The length of the label is equal to the soup can's circumference (C) plus 1/4" for the overlap.

$$C = \pi \times D = \pi \times 2\,^5/_8 = 3.14 \times (2.625) = 8.24 \text{ inches}$$

$$\text{Label length} = C + \,^1/_4" = 8.24 + 0.25 = 8.49$$

$$\text{Label length} = 8.50 \text{ inches (rounded)}$$

Thus the label for each can is 8.5 inches long.

The paper for printing the labels is purchased in rolls 3 feet wide and 1000 feet long. The paper is printed and then cut with large machines that have sharp cutting rollers. Along which direction should the labels be printed to get the most labels per roll of paper with the least amount of waste? How many labels can you print per roll?

To determine the best direction to print the labels, you must calculate the number of labels that can be printed with each arrangement.

Look at the two pictures shown below. One shows labels arranged horizontally along the 1000-ft rolls. The other shows the labels arranged vertically.

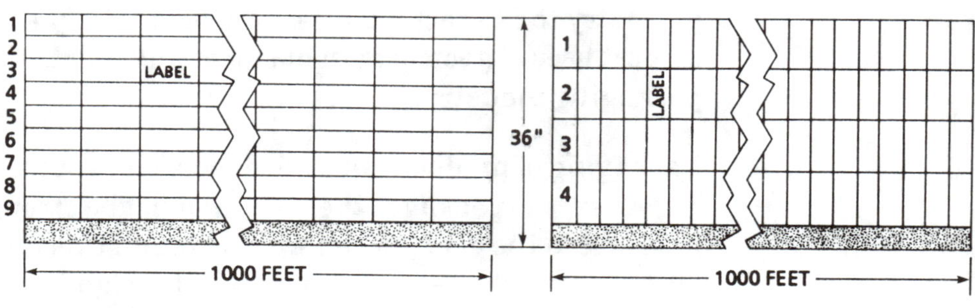

Using the dimensions of the label from Example 8, determine which geometry gives the larger number of labels (least waste) per 1000-ft roll.

Creating circle charts

You work in the accounting department of a department store. You are asked to draw a "pie" chart of the distribution of sales. Your supervisor gives you the following information:

> 58% of sales is for merchandise
> 12% of sales is selling expense
> 10% of sales is administrative expense
> 4% of sales is interest on borrowed money
> 6% of sales is paid toward income taxes
> 2% of sales is for delivery charges
> 7% of sales is retained as profit
> 1% of sales is for miscellaneous expenses

The total adds to 100%—as it should.

The easiest way to make this chart is to calculate the interior angles of each section of the "pie." We'll round each answer to the nearer whole degree, making sure that the sum of angles equals 360°.

$$58/100(360) = 208.8 \text{ degrees} = 209 \text{ degrees}$$
$$12/100(360) = 43.2 \text{ degrees} = 43 \text{ degrees}$$
$$10/100(360) = 36.0 \text{ degrees} = 36 \text{ degrees}$$
$$4/100(360) = 14.4 \text{ degrees} = 14 \text{ degrees}$$
$$6/100(360) = 21.6 \text{ degrees} = 22 \text{ degrees}$$
$$2/100(360) = 7.2 \text{ degrees} = 7 \text{ degrees}$$
$$7/100(360) = 25.2 \text{ degrees} = 25 \text{ degrees}$$
$$1/100(360) = 3.6 \text{ degrees} = 4 \text{ degrees}$$

Adding up the angles verifies that rounding yields a sum of 360°. So you should draw a circle, and then use a protractor to divide it into the sections, as shown below.

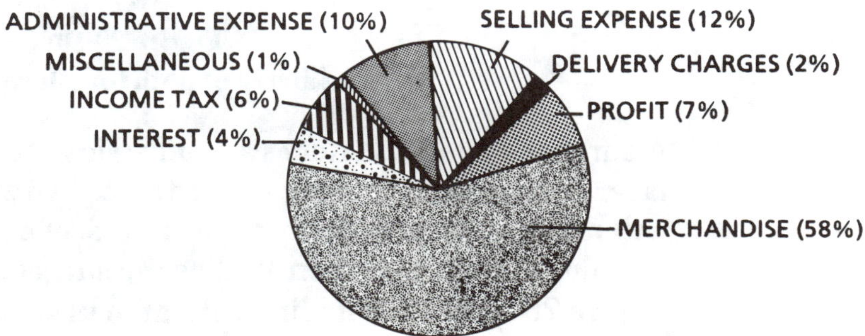

GEOMETRY IN HEALTH OCCUPATIONS

Health occupations encompass a wide variety of job activities and descriptions. Modern hospitals have more than just facilities to heal the sick. In many cases, people who are being treated stay overnight, or for even longer periods of time.

In many ways, a hospital can be considered to be a "home" away from home. That is, a modern hospital offers almost everything patients have at home—and more. In addition, a hospital has stores, shopping areas, and restaurants. Since a modern hospital is open 24 hours a day, 365 days a year, they have three work shifts—and thus three people for every job.

Geometry becomes involved in many aspects of the design and layout of hospital facilities, design of special beds, physical-therapy devices, and so on. We will show a typical need for geometry in the hospital environment—one that involves laser technology and surgeons.

Laser technology is a rapidly growing area in the medical field. A laser emits a high-intensity, focused beam of light to cut, burn, or heat specific areas of tissue. To reduce human error in applying the laser beam to the appropriate area, a *robotic arm* is used to hold and guide the laser through a preprogrammed series of movements.

The desired pattern of movement (see Figure 28-10) is carefully determined and programmed into a computer. When the laser is to be used, the computer takes over and guides the robotic arm, pointing the laser beam at the correct position for the correct amount of time.

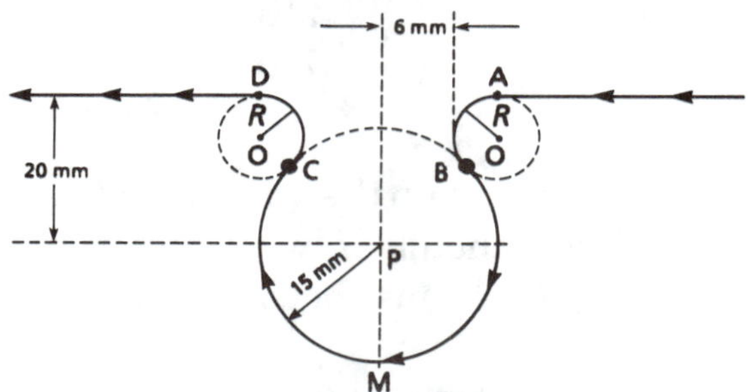

Figure 28-10
Geometry of path for a laser beam

A surgeon usually sketches a geometric pattern that will guide the laser beam around the area to be treated with the laser. A surgeon will make a careful sketch, for example, of a patient's heart valve. A computer technician then studies the surgeon's sketch—shown in Figure 28-10. The solid line with arrows shows the programmed path the laser beam should follow. Before the technician can program the information into the computer, the radius R of the identical arc paths **AB** and **CD** must be determined.

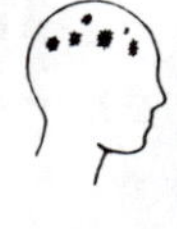

Study the diagram in Figure 28-10 carefully and note the following:

- The path begins at the upper right as a straight line. It then changes successively and smoothly into the arc of a circle of radius R, then to a larger circle of radius 15 mm, next to another arc of radius R, and finally moves off to the left as another straight line.

- The two straight-line segments are tangent to the small circles at points **A** and **D**.

- The two small circles are tangent to the large circle of radius 15 mm at points **B** and **C**.

- The two straight lines are 20 mm above the center of the large circle.

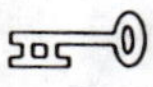

To find the unknown dimension R, redraw Figure 28-10 as shown in Figure 28-11, with right triangle **OPQ** clearly defined. Study both diagrams.

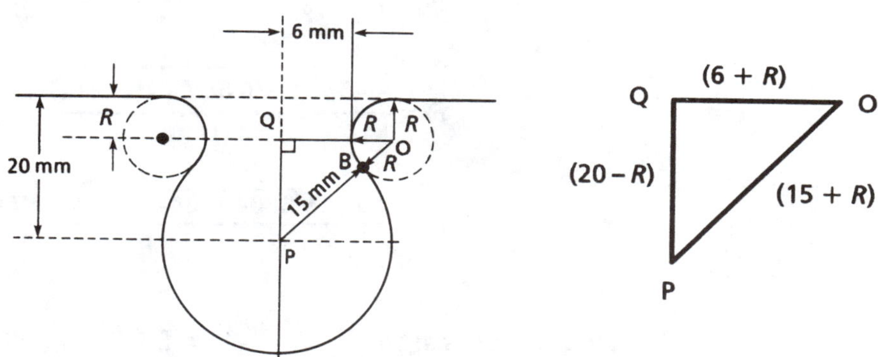

Figure 28-11
Geometry in ΔOPQ to determine R

Note the following:

- Line segment **OP** passes through a common point of tangency (point B in Figure 28-10) so that the length of **OP** is equal to $(15 + R)$.

- Line segment **OQ** forms a right angle with line segment **PQ** at point **Q**. That is, $\angle$**OQP** $= 90°$.

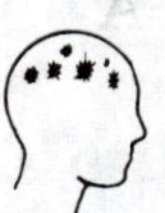

- Line segment **OQ** is equal in length to $(6 + R)$.

- Line segment **PQ** is equal in length to $(20 - R)$.

- Since $\triangle$**OPQ** is a right triangle, the Pythagorean theorem tells us that $(\mathbf{PQ})^2 + (\mathbf{OQ})^2 = (\mathbf{OP})^2$.

Substituting for **PQ**, **OQ**, and **OP** in the Pythagorean theorem gives:

$$(20 - R)^2 + (6 + R)^2 = (15 + R)^2$$

Squaring the three binomial factors—a skill you learned in Unit 23, *Factoring*—you get:

$$(400 - 40R + R^2) + (36 + 12R + R^2) = (225 + 30R + R^2)$$

Collecting terms in R^2, R, and constants:

$$(R^2 + R^2 - R^2) + (-40R + 12R - 30R) + (400 + 36 - 225) = 0$$

Simplifying the terms in each of the parentheses, you get:

$$R^2 - 58R + 211 = 0$$

Solve for R with the quadratic formula,

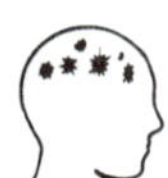

$$R = \frac{-b \pm \sqrt{b^2 - 4ac}}{2a} \quad \text{where } a = 1, b = -58, c = 211$$

$$R = \frac{-(-58) \pm \sqrt{(-58)^2 - 4(1)(211)}}{2(1)}$$

$$R = \frac{58 \pm \sqrt{3364 - 844}}{2} = \frac{58 \pm 50.2}{2}$$

Taking the *plus* sign, $R = \dfrac{108.2}{2} = 54.1$ mm. (This answer does not fit the problem. Why not?)

Taking the *negative* sign, $R = \dfrac{7.8}{2} = 3.9$ mm. (This answer fits the problem. Thus $R = 3.9$ mm.)

Having found the unknown dimension R, the computer technician is now able to program the correct path for the robotic arm. Thus, the laser beam will travel along the heart valve boundaries as **required by the surgeon.**

Based on what you've learned in Example 10, solve for the unknown radius R in the geometrical path (follow the arrows) shown below. The points **B** and **C** are *points of tangency* between the large circle and the two small circles. As in Example 10, draw the right triangle that involves the unknown radius R. That will help you solve for the value R.

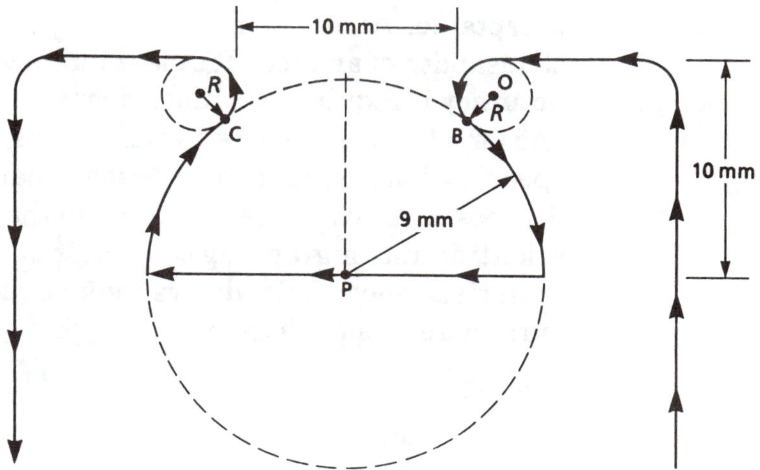

SUMMARY

This unit has shown you how *two-dimensional geometry* is used to solve problems around the home and in the workplace—in industry, agriculture, marketing, economics and hospitals. The principles of geometry that you applied to solve problems included:

- Areas of rectangles, triangles, trapezoids, circles in various arrangements.

- Sectors of circles and segments of circles.

- Circumference of circles.

- Relationship between a line tangent to a circle and a radius drawn to the point of tangency.

- The characteristics of special triangles (equilateral, right, isosceles, 30°–60°–90°, 45°–45°–90° and 3–4–5).

- The Pythagorean theorem.

- Line segments that bisect angles and other line segments.

- Sine, cosine and tangent functions that relate sides and angles in right triangles.

You learned that calculating lengths, areas and angles is necessary for almost every problem that involves two-dimensional geometry.

You learned also that often you have to draw *auxiliary diagrams* to help you find an unknown dimension or angle.

And finally, you learned that every problem in the workplace doesn't have to be calculated to the same *degree of accuracy*. Sometimes a rough estimate—to the nearest foot or inch, for example—is acceptable. At other times you may need answers to the nearest ten thousandth of an inch. It's clear for instance, that a higher degree of accuracy is required to build a door for the space shuttle than to build one for a barn. In general, whenever high cost of materials or tight specifications are involved, measurements and calculations to higher degrees of accuracy are required. In these instances, the use of a scientific and/or graphing calculator is very helpful. When you buy materials, you should always *round up* from your estimates, just to be sure there is enough to finish the job.

Laboratory Activities

Use the mathematics skills you have learned to complete one or all of the following activities.

Activity 1   **Measuring distances with a cylinder**

Equipment

Cloth measuring tape
Tape measure, English or metric
Three-pound coffee can
One-pound coffee can
One-inch wooden dowel
Vernier caliper
Masking tape
Calculator
Felt-tip marker

Statement of Problem

Most of the Egyptian pyramids we know about today were built between 2815 B.C. and 2294 B.C. One of the major problems with building the pyramids was the need for accurate measurements. Because measuring devices of this era could stretch or shrink, depending on the weather or how tightly the device was pulled, accurate measurements were not easy to make. To solve this problem the Egyptians measured the circumference of a large cylinder. Then they would roll the cylinder and count the number of revolutions needed to cover a distance.

In this activity you will measure distance using cylinders of different diameters.

Procedure

a. Use the vernier caliper, cloth tape measure, or steel tape measure to determine accurately the **outside** circumference of each of the cylinders—the two cans and the dowel rod. You'll need to decide whether it is more accurate to measure the circumference with the cloth tape, or to measure the diameter with other instruments and then calculate the circumference. Record the measurements in your Data Table (see sample on following page).

b. Wrap a single strip of masking tape around the circumference of each cylinder. Using the marker, make *sixteen* equally spaced marks on the masking tape. Do this by first marking halves, then quarters, then eighths, then sixteenths. Make one of the marks larger, so that you will recognize one complete revolution of the cylinder.

c. Measure out a distance of 300 cm with the steel tape measure on the floor of your room (or 100 inches if you are working in the English system). Mark this distance accurately with two pieces of masking tape—one at the start and one at the finish.

d. Position the largest cylinder with its guide-mark "touching" the **start** line. Carefully and slowly roll it toward the "finish line," counting the complete turns as you go. When you near the **finish** line, count the number of divisions on the tape between the last full revolution and the finish line. Determine how many complete revolutions and sixteenths of a complete revolution you've made. Record the total count in your Data Table, to the nearest sixteenth of a revolution.

e. Repeat Step d two more times, for a total of three measurements for the cylinder. Average the three trials, and record the average number of revolutions in the Data Table for this cylinder.

f. Repeat Steps d and e for the other two cylinders.

g. Measure out a distance of 30 cm (or 12 inches if you're using the English system) with the steel tape measure, and mark the start and finish with masking tape.

h. Repeat Steps d and e for all three cylinders, measuring the 30-cm distance.

DATA TABLE

Cylinder	Diameter	Circumf.	Trial 1	Trial 2	Trial 3	Avg.	Calculated Distance
Large Can							
Small Can							
Dowel							

a. For cylinders where you *measured the diameter*, calculate the circumference for each cylinder. Record the circumference value in your Data Table.

b. Using the average number of revolutions needed to cover each of the distances, determine the total distance rolled for each cylinder. Record these values in your Data Table.

c. Which of the cylinders performed best in measuring the long distance? the short distance? Explain these results.

d. What size cylinder would you recommend if you wanted to accurately measure the length of a football field—100 yards? Explain your answer.

e. What size cylinder do you think the Egyptians used to construct the pyramids?

f. In Step b of the *Procedure*, you divided the circumference into sixteenths. Why not divide it into sevenths, or thirds, or seventeenths? Would your results have been any different if you had marked the tape with differently spaced divisions?

Activity 2   Surveying and the law of sines

Equipment

Construction paper (or card stock)
Masking tape
Protractor
Tape measure, English or metric
Calculator with trigonometric functions

Statement of Problem

Surveyors often need to measure the length or span of a distance, even though they can't actually stretch a tape measure across the distance. You may have performed the activity of Unit 22, where we used the *law of cosines* to find an unknown distance when one angle and two sides of any triangle were known. In this activity, you'll use the *law of cosines* and *law of sines* to find an unknown distance when you know one side and two angles of a triangle.

For $\triangle\mathbf{ABC}$ with sides and angles as shown in the following figure, the relationship given by the law of cosines and the law of sines is as follows.

Law of cosines

$$c^2 = a^2 + b^2 - 2ab \cos \mathbf{C}$$
$$a^2 = b^2 + c^2 - 2bc \cos \mathbf{A}$$
$$b^2 = a^2 + c^2 - 2ac \cos \mathbf{B}$$

Law of sines

$$\frac{\sin \mathbf{A}}{a} = \frac{\sin \mathbf{B}}{b} = \frac{\sin \mathbf{C}}{c}$$

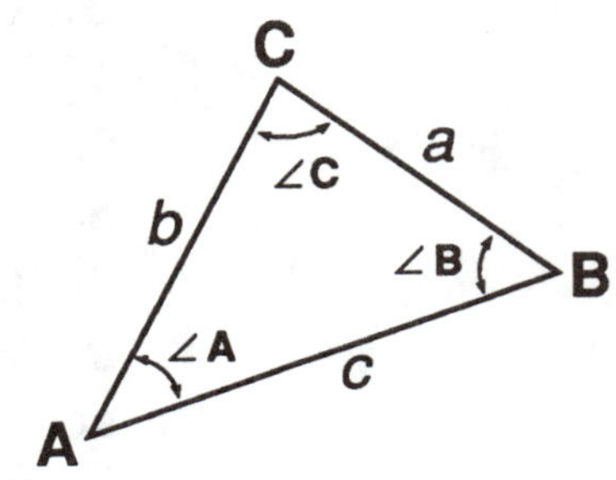

For example, in $\triangle\mathbf{ABC}$, suppose you know angles $\mathbf{A}$ and $\mathbf{B}$, and the side a opposite angle $\mathbf{A}$. Using the law of sines, you can find the length of side b opposite angle $\mathbf{B}$. In this activity, you will determine the size of a distant object by measuring two angles and a baseline distance.

Procedure

a. Your teacher will tell you what object will be measured. It might be an object in your classroom, an object outside a window, or even an object that is pretty far away from your classroom, for example, across a parking lot. The following directions will assume that you are measuring an object across the classroom—an edge of a long desk, for example.

b. You will be measuring four angles, as depicted in the drawing here: $\angle\mathbf{1}$, $\angle\mathbf{2}$, $\angle\mathbf{3}$, and $\angle\mathbf{4}$.

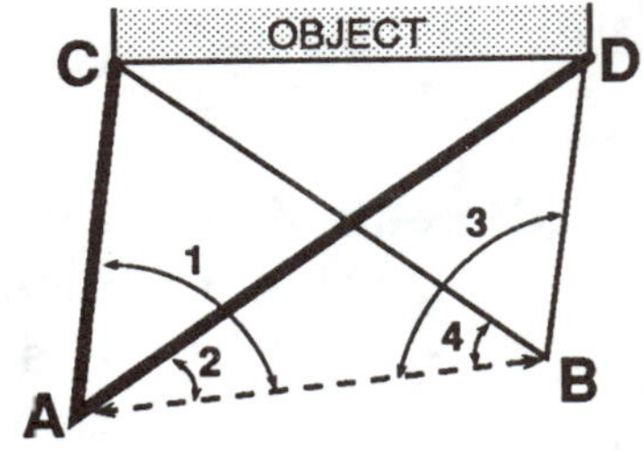

- Identify two points, **C** and **D**, on the object to be measured, as shown in the drawing here. Your goal will be to determine the length **CD** by using the law of sines and the law of cosines.

- Determine the two points, **A** and **B**, from which you'll make your angle measurements. (Notice that the line **AB** need not be parallel to the line **CD**. You teacher will give you some guidance here.) Place a small piece of masking tape where points **A** and **B** are located. You may find it helpful to locate points **A** and **B** on a tabletop where it will be easier to measure angles with a protractor.

- Measure the distance **AB** with the tape measure and record it in the Data Table. (See next page.)

CORD Applied Mathematics

- Notice that you will measure $\angle 1$ and $\angle 2$ from point **A** (see the bold lines drawn from point **A**). And you will measure $\angle 3$ and $\angle 4$ from point **B** (see the thin lines drawn from point **B**). It is very important that all of these angles be measured with respect to line **AB**, as discussed in the next step.

c. Measure the angles, beginning at point **A**. Align the *baseline of your protractor* with the baseline **AB**, as shown in the drawing here. (Ask your teacher for help, if necessary.) Position the center point of the protractor directly over **A**. Determine angle **1**, as shown in the drawing, to the nearest 0.5°. You can use the construction paper (or card stock) to help you sight from **A** to **C**. In a similar fashion, determine angle **2**, sighting from **A** to **D**. Record the angle measurements in the Data Table.

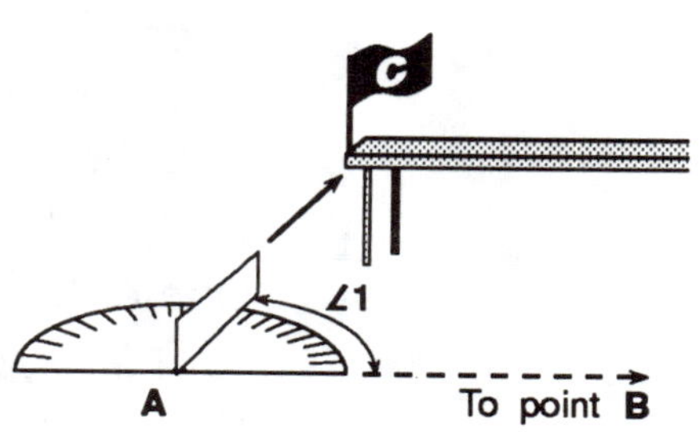

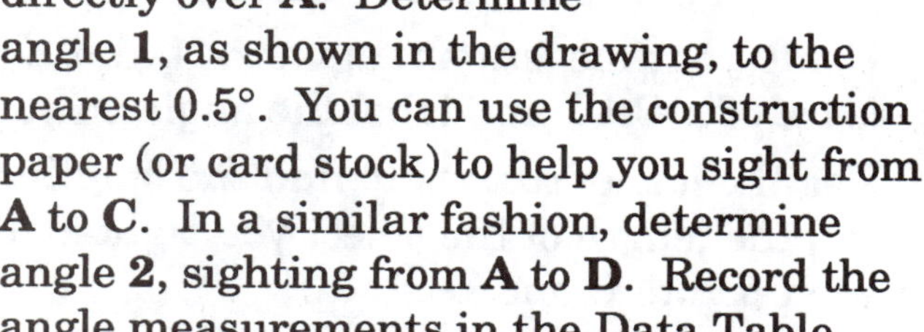

Data Table

Length **AB** =
$\angle 1$ =
$\angle 2$ =
$\angle 3$ =
$\angle 4$ =

d. Now, move to Point **B**. Align the protractor's baseline with the line **AB**, as before. Sight from **B** to **D** and measure angle **3**. Then sight from **B** to **C** and measure angle **4**. Record the angle measurements in your Data Table.

Calculations

Remember, your goal is to determine the length **CD**. At present you know only the length **AB**. You also know several angles ($\angle 1$, $\angle 2$, $\angle 3$, and $\angle 4$) that belong to $\triangle ABC$ and $\triangle ABD$ that include side **AB**. You should remember three useful formulas when working with triangles:

1. The angles within a triangle sum to 180°.

2. The *law of sines* gives the relationship between a pair of angles and the sides opposite them. (See the Statement of the Problem at the start of this lab.)

3. The *law of cosines* gives the relationship between an unknown side and the angle opposite that side, if the length of the two other sides is known. (See the Statement of the Problem at the start of this lab.)

a. Identify the two overlapping triangles that include the known side **AB**: $\triangle ABC$ and $\triangle ABD$. (Refer to the figure beside Step b.)

Draw a sketch of each on your paper, showing the angles and sides that you know. Use the values in your Data Table.

b. Determine *all* the angle measurements for Δ**ABC** and Δ**ABD**. Label the angles on your sketches.

c. Use the *law of sines* with Δ**ABC** to find the length of side **AC**. Label the length of side **AC** on your sketch. (Hint: Select pairs of sides and angles to substitute into the *law of sines*. Solve for the unknown side **AC**.)

d. Similarly, use the *law of sines* with Δ**ABD** to find the length of side **AD**. Label the length of side **AD** on your sketch.

e. Make a sketch of Δ**ACD**. (Refer to the drawing beside Procedure Step b.) Label the known sides **AC** and **AD** that you determined in Steps c and d above.

f. Using your measurements for ∠**1** and ∠**2**, determine the measure of ∠**CAD** and label the angle on your sketch of Δ**ACD**.

g. Use the *law of cosines* to find the length of side **CD** in Δ**ACD**, that is, the length of the object you sighted (or the distance between points **C** and **D**).

h. If possible, use your tape measure to actually measure the length of the object. Compare the results of your calculation with the actual length. Do they agree? If not, what explanation can you suggest?

Activity 3	### Resurfacing a parking lot
Equipment	Measuring tape String or cord Protractor Drawing kit (Accu-Line™) Calculator
Statement of Problem	A business in town wants to repave its parking lot. You are asked to obtain an estimate of how much asphalt will be needed. The construction company says that they need to know how many square feet of area the parking lot covers.
Procedure	Your teacher will supply some guidelines for this activity. Weather conditions permitting, you will make a sketch of your school's parking lot surface, as directed by your teacher. (If this is not possible, your teacher will give you an alternate surface to

measure.) You will then make the measurements needed to determine the total area to be paved, in square feet or square meters.

a. Take careful notes as your teacher provides the instructions for
the area to be paved. The instructions may include landmarks,
distances, angles, etc.

b. Use your drawing kit to make a sketch of the area to be paved.
Identify geometric shapes that make up the total area. You will
use the area of these individual shapes to determine the total area.
(Note: There may be more than one way to do this. Consider
different ways so that you can find the way that might be the
easiest.)

As an example, consider the
sketch of a certain lot shown
here. The shaded portion is
the area to be paved. The
dotted lines show how it can
be broken into rectangles, triangles, and
portions of circles. This is just one of
many ways that *this lot* could be divided.

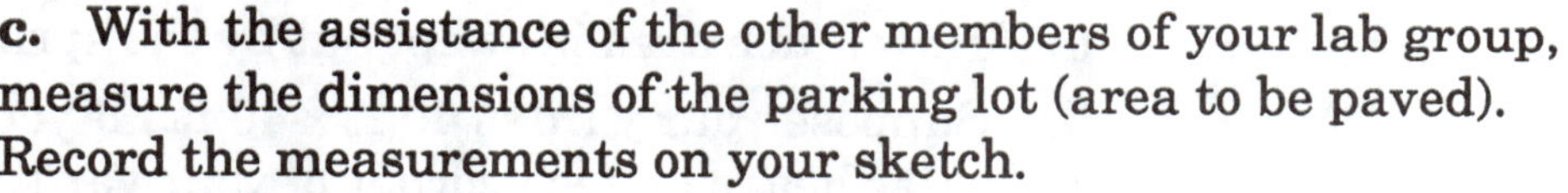

c. With the assistance of the other members of your lab group,
measure the dimensions of the parking lot (area to be paved).
Record the measurements on your sketch.

Be sure to record the total lengths of rectangular regions, as well as smaller subsections.
For circular regions that are not complete
circles, you may need to determine the angu-
lar measure. One way to do this is to use
string to locate the *center* of the arc. The

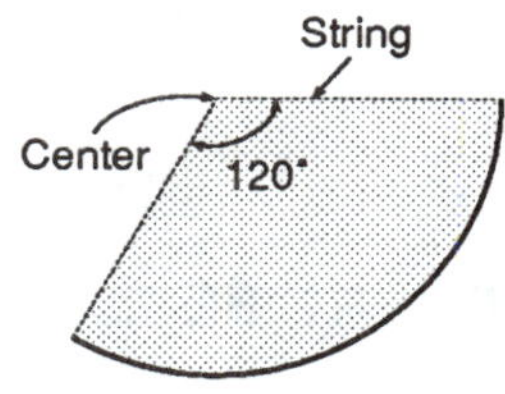

distance from the center to the arc is the *radius* of a circle. Then
use your protractor to measure the *angle* made by the two sides, as
shown in the drawing here. If the region has an angle of 360°,
then its area would be equal to the area enclosed by the circle. If
it is only 120°, then it would be $^{120}\!/_{360}$, or $\frac{1}{3}$ the area of the circle.

<table>
<tr><td>Calculations</td><td>

When you're finished measuring, return to the classroom to
complete the area calculations. If you've already identified the
geometric shapes and the area formulas, your calculations should
be rather simple. You can write the area for each piece on your
sketch. Then compute the total area by summing the area of each
piece. If other groups in the class worked with the same area,
compare your group's results with that of the other groups'.
</td></tr>
</table>

Student Exercises

You can solve the exercises that follow by applying the mathematics skills you've learned. The problems described here are those you may meet in the world of work.

Note: *Wherever possible, use your calculator to solve the problems that require numerical answers.*

GENERAL

EXERCISE 1:

You are installing a sprinkler system to water your yard. The pulsating sprinkler head you've chosen will spray water a distance of 25 ft. You position one head at the corner of your yard, covering a circular wedge over an angle of 90 degrees, similar to that shown here.

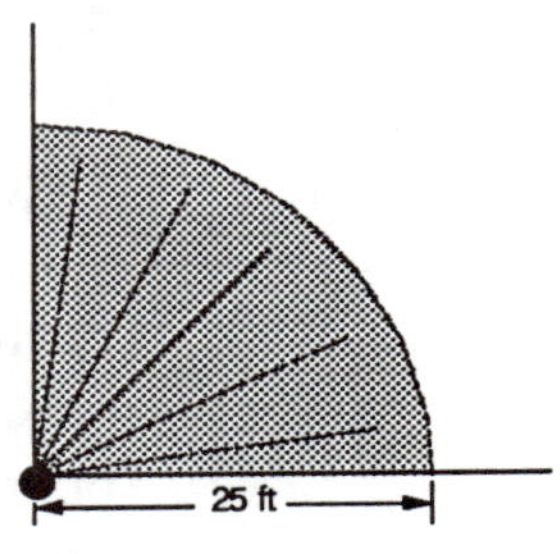

a. How much area is swept out by the sprinkler?

b. Suppose your yard is 40 ft × 40 ft. If you placed one of these sprinkler heads at each corner of your yard, would you be able to cover all the yard? If not, how far would each of the four corner sprinkler heads need to reach to cover all the yard? Would there be any "overlap" regions?

EXERCISE 2:

You are piloting a sightseeing boat carrying press photographers for a sailboat race. The race course first heads the boats in a direction *38° east of north* for 1.5 nautical miles to the first waypoint. The boats then must turn to *37 degrees west of north* for 0.8 nautical mile to the second waypoint. Finally, the boats return to the starting point. (See the drawing here.) You want to find the course necessary to get to the *second waypoint* from the starting point. There is a formula called the *law of cosines* that can help you: $c^2 = a^2 + b^2 - 2ab \cos \mathbf{C}$. For any triangle ($\triangle \mathbf{ABC}$), you can compute the length of the third side of a triangle (c), if you know

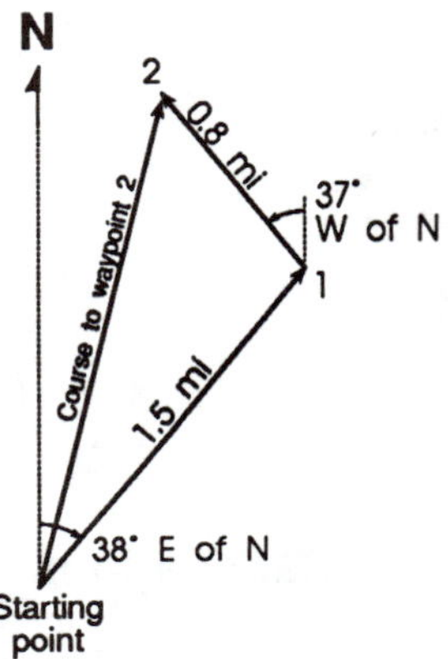

the lengths of the other two sides (*a* and *b*) and the angle between them ($\angle$C). And in a similar way, there is the *law of sines*:

$$\frac{\sin \mathbf{A}}{a} = \frac{\sin \mathbf{B}}{b} = \frac{\sin \mathbf{C}}{c}$$

This law helps you find an angle in a triangle if you know the length of the side that's opposite it (and another side and angle).

a. Identify the triangle depicted in the drawing by the race course. Make a sketch on your paper of the triangle. Label the vertices and sides of your triangle, such that the two sides of known length are labeled *a* and *b*.

b. Determine the angle between side *a* and side *b*, that is, $\angle$C.

c. Use the *law of cosines* to find the distance from the starting point to waypoint 2.

d. Use the *law of sines* to find the bearing angle (that is, how many degrees east of north) for the sightseeing boat's course from the starting point to waypoint 2.

EXERCISE 3: An airplane leaves its home airport heading 38° east of north. It flies for 96 miles on this course and then turns to 90° (due east) and flies for another 120 miles. Make a sketch of the *triangle* involved in this problem. If the airplane's radio signal has a range of about 200 miles, is the plane still within radio range of the home airport? (Hint: See the *law of cosines* given in Exercise 2.)

AGRICULTURE AND AGRIBUSINESS

EXERCISE 4: You are surveying your land for the placement of poles to support a power line to a remote barn. You need to have the lines cross a river. You don't want to have a single span of wire longer than 300 ft. You sight across the pond to the proposed pole position. Then you move a perpendicular distance of 50 feet and resight to the same position. The angle of the second sighting relative to the first is 7.4°, as shown here (not to scale). Is the distance across the lake (*d*) more than the maximum distance?

EXERCISE 5:

You are hired to plant bushes around the perimeter of the grassy sections in a parking lot. Each grassy section has the shape of a rectangle with half-circles on the ends, as shown below. The bushes you will plant need 3 feet of growing room. This means you must allow space both between adjacent plants and between the bush and the curb. Thus you need to determine the *planting path*, as shown in the drawing.

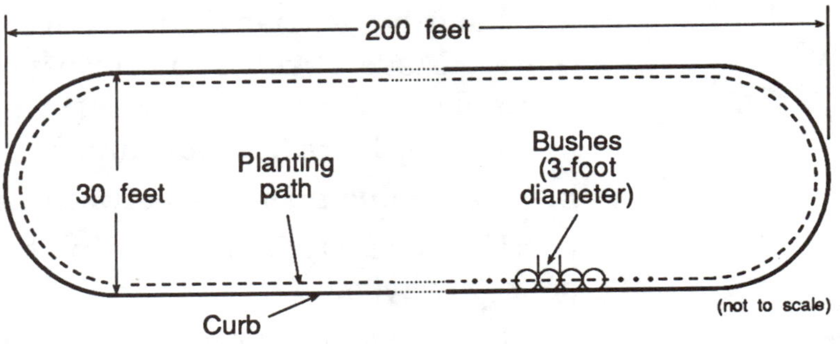

a. What is the perimeter of the planting path? (Hint: Remember to allow for the growing room of each bush.)

b. How many bushes will be required for each grassy section? (There will be a minor complication caused by the placement of the bushes around *circular* paths—ignore this.)

EXERCISE 6:

You need to seed and fertilize the infield of a half-mile horse track. (The length of a running track is normally based on the distance around the inside rail.) The track—semicircular ends joined by a straightaway—fits in a field that is 1200 feet long, as shown here. The width of the running area is 35 feet. To order the proper amounts of seed and fertilizer, you need to determine how many acres are in the *infield*. (Note: 1 acre = 43,560 ft^2; 1 mile = 5280 feet.)

HALF-MILE TRACK

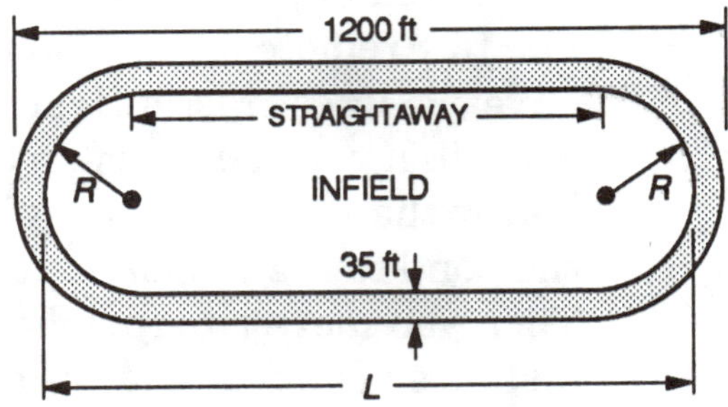

a. What is the length L of the infield? (See drawing.)

b. What is the total distance P around the inside rail, in feet?

CORD Applied Mathematics

c. With L as the total length of the infield and R as the radius of the turns, write an expression for the length of the straightaway sections, S.

d. Similarly, write an expression for the inside distance around each turn, T. That is, determine the distance around the inside rail for each semicircle.

e. Write an expression for the total distance P around the infield, using the results of Parts c and d.

f. Substitute the known values for P and L and solve for the radius of the turns, R.

g. Using a sketch of the infield area, identify rectangular and circular (or semicircular) regions. Compute the total area of the infield using the dimensions you've determined above, and convert it to acres.

EXERCISE 7: You have been hired to set up an irrigation system on a farm. The field to be irrigated is a *parallelogram* 3000 feet wide and 3500 feet long with an angle of 65 degrees. The farmer wants to use circular irrigation units that cover a circular area (or a part of a circular area) of 500-foot radius. Find the minimum number of irrigation units that will cover as much area as possible.

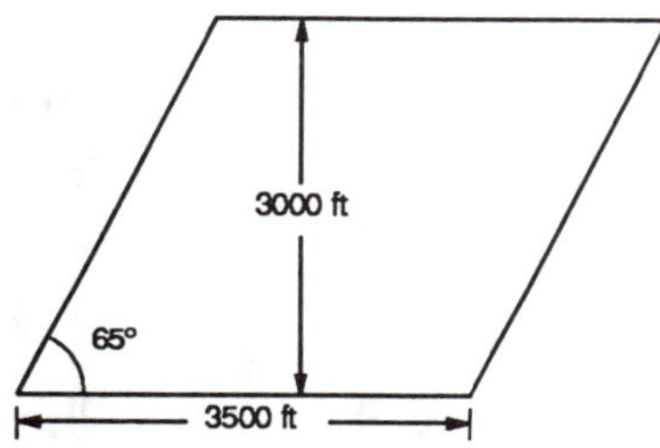

a. Find the total area of the field.

b. Find the area that each circular irrigation unit can possibly cover.

c. Find the number of units that would be needed ideally, if each unit could, in fact, cover its maximum area (a circular region).

d. Make a scale drawing of the land and the area covered by an irrigation unit. If no water is to be sprayed outside the boundaries of the field, how many units do you think are actually needed? Compare this answer with the answer from Part c. (Hint: There is more than one correct answer.)

BUSINESS AND MARKETING

EXERCISE 8: You work in the accounting department of a wholesale company. For an upcoming meeting, you need to present the following sales data:

Item	% of total sales
Office equipment	17%
Home appliances	5%
Clothing	20%
Sporting equipment	35%
Yard equipment	3%
Computer supplies	10%
Computers	10%

Prepare a "pie chart" to represent this data. Show your work in determining the size of each section. (If you have access to a spreadsheet/graphing program, also enter the data and let the program create a "pie chart.")

EXERCISE 9: The parking lot for your business was recently repaved, so you need to repaint the stripes. Each space is to be 9 feet wide and 18 feet long, angled at 30 degrees, as shown here.

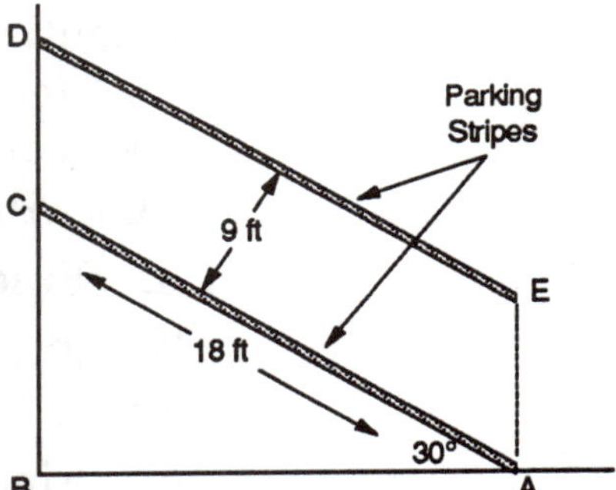

a. The painter needs to know where to start the line for the corner parking space. That is, the painter needs the distance **AB**, from the corner to point **A**. What is the distance **AB**?

b. You need to tell the painter the point where the first line (that begins at **A**) will end. Find the distance from point **B**, the corner, to point **C** (length **BC**).

c. Find the distance between consecutive painted lines, that is, the length **CD** (or **AE**).

CORD Applied Mathematics

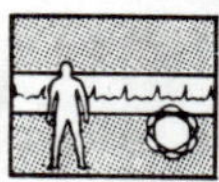

HEALTH OCCUPATIONS

EXERCISE 10:

One of the problems with artificial joints for the body is maintaining the same flexibility as the natural joint. This is particularly difficult with knee joints. If the length from the knee joint down to the foot is 20 inches and the foot must be able to move sideways from right to left by 0.75 inch, find the *total angle* that the knee must be able to flex, from side to side.

EXERCISE 11:

Replacement parts for the body are normally made by computer-controlled machines. The doctor provides various dimensions, and the computer technician must give the proper dimensions to the computer program. A particular joint might look similar to that shown here, with a few labeled dimensions. In

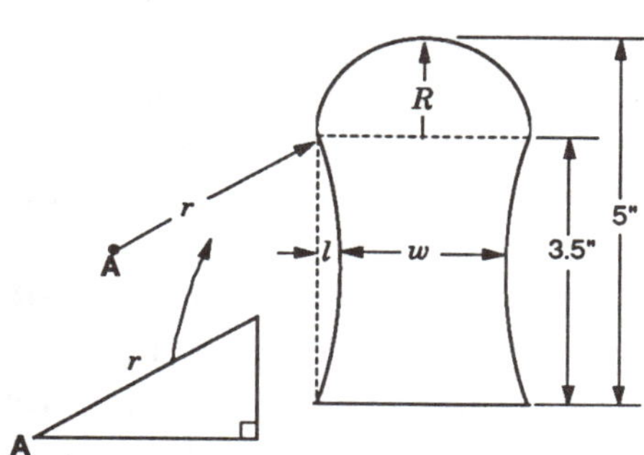

this case, the overall height must be 5 inches, the height of the lower "socket" (less the radius of the hemispherical knuckle) is 3.5 inches, and the radius of curvature of the lower "socket" (r) is 6 inches. You need to find R, l, and w. Let's do this in several steps.

a. Using the given dimensions, find the radius R of the hemispherical knuckle on top of the joint.

b. To find l, notice the right triangle shown beside the drawing. Identify the dimensions of the right triangle, in terms of l and the given dimensions.

c. Use the Pythagorean formula to write an expression that includes the unknown dimension, l, the side of the right triangle. Expand the formula to obtain a quadratic equation involving l.

d. Use the quadratic formula to determine the value of l that satisfies your equation from Part c.

e. Draw a second vertical dotted line on the right side of the lower socket, like the one drawn on the left. Determine the dimension w, using the value for l that you determined and the distance between the vertical dotted lines.

HOME ECONOMICS

EXERCISE 12:
Part of a sewing project requires a section of cloth with a taper of 0.9 inch per foot. (The *taper* of an object is the difference between the beginning and ending widths. The amount of taper is often given as a fraction of an inch per foot of length as, for example, is the case here: 0.9 inch per foot.) The beginning width is 1.5 feet and the ending width is 1.2 feet, as shown in the drawing. The length referred to here is measured along the center line of the piece of work and is labeled L in the drawing.

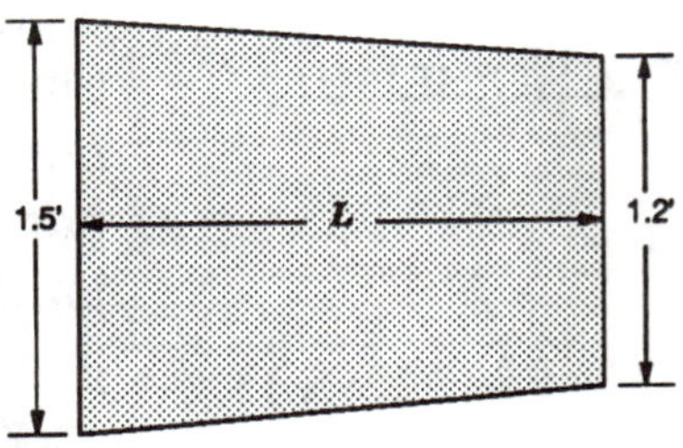

a. Express the given taper of 0.9 inch (of width) per foot (of length) as a ratio. Set up a second taper ratio: the total change in width for the section of cloth divided by the length L. Equate these two ratios as a proportion and solve for L.

b. Use the dimensions given and the length (L) to find the area of the piece of cloth.

EXERCISE 13:
Your company manufactures cloth napkins. You want to lower the manufacturing cost by reducing waste. One way to do this is to reduce the amount of cloth wasted when making the napkins. Each napkin requires 12 inches by 16 inches of material. Each bolt of cloth is 60 inches wide and 25 yards long. The drawings below show two proposed ways to cut the napkin material. To minimize the amount of waste per bolt of cloth, which of the two orientations is better?

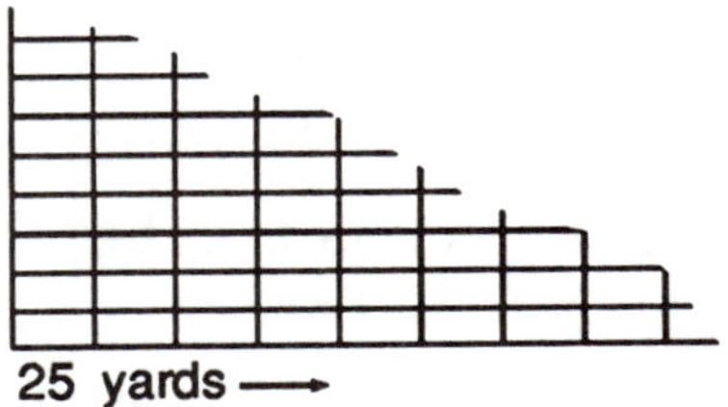

a. Orientation 1

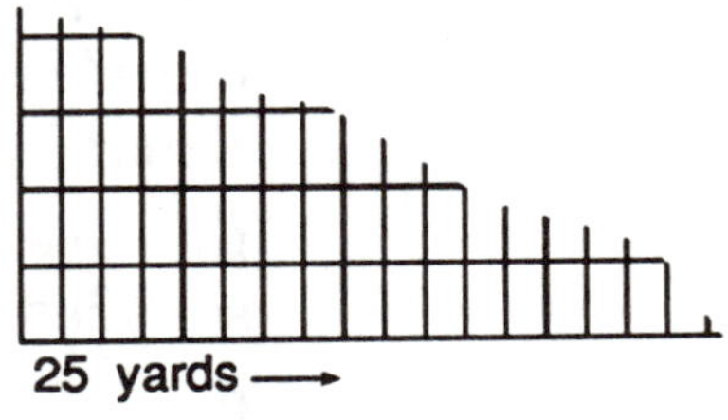

b. Orientation 2

CORD Applied Mathematics

EXERCISE 14:   The lamp housing of a simple spotlight forms a cone of light with an angle of divergence of 40°, as shown in the drawing. The lamp is mounted 10 ft above the floor. Assume that the light is a point source, as shown. To determine the intensity of the light at this height, you need to determine the area of the circular spot on the floor.

a.   Identify the triangle formed by the side view of the cone of the light. Bisect the triangle by dropping a perpendicular line from the light to the floor. Identify the known angles and height in the triangles.

b.   Calculate the length of the base of either of the right triangles. This is the radius of the spot on the floor, r.

c.   Calculate the area of the circular spot on the floor.

d.   If the light were raised to 12 feet, what would be the area of the spot on the floor?

EXERCISE 15:   You are selecting a television set to fit in your entertainment cabinet. The cabinet shelf is 24 inches wide. You want to allow 2 inches on either side for the television housing. What is the maximum diagonal dimension of the television screen possible, if the aspect ratio of the screen (that is, height to width ratio) is 3:4?

EXERCISE 16:   You are setting up a new stereo system in your home. You want to find the best spot in the room for listening. The "face centers" of the two speakers are separated by a distance of 20 feet. Each speaker is "turned in" 25°, as shown in the

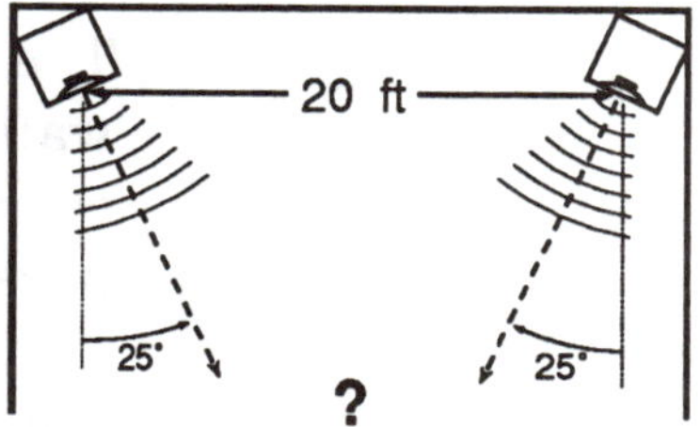

drawing here. You want to find where you should place your chair, so that the sound from the speakers is directed toward the chair and is equally balanced.

a.   Draw a triangle on your paper that represents the geometry of the problem. Label the known dimension and angles.

b.   For balanced sound, the chair must be equally distant from both speakers. How far from each speaker must the chair be positioned? (Hint: Draw a line from the chair, perpendicular to the wall between the speakers, bisecting the triangle.)

c. Suppose you want to angle the speakers so that the optimum location is 12 feet from the speakers. At what angle should the speakers be positioned?

INDUSTRIAL TECHNOLOGY

EXERCISE 17: You have been hired to paint a large church wall that is 60 feet long and 16 feet high. The wall has 7 unusually shaped windows. Each window has a set of panes in the shape of a half-circle on top of a rectangular window that is 8 ft high and 4 ft wide, as shown here. To estimate the amount of paint needed, you need to know the total area of the surface to be painted.

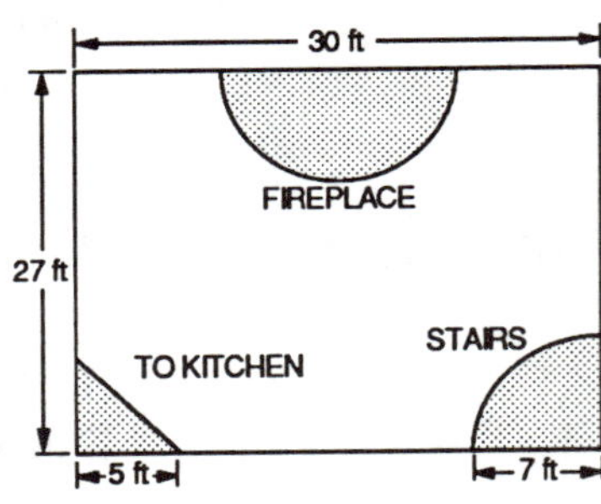

a. Determine the area of the entire wall, including the windows.

b. Determine the area of each window by finding the area of separate parts. (Hint: What is the radius of the circular portion of the window?)

c. Subtract the area of all the windows from the total wall area to determine the area to be painted.

EXERCISE 18: You are going to put down tile in a friend's family room. The family room is 27 ft by 30 ft, but there are several areas that will not be tiled. The room has a fireplace on one wall, an exit to the kitchen, and an entrance from the stairs, as shown here. The fireplace cuts out an area that is a half-circle of radius 7 ft.

The entrance from the stair is a quarter-circle, also of radius 7 ft. The exit to the kitchen cuts out a triangular section in the corner of the room, with each leg measuring 5 ft.

a. Find the overall area of the 27' × 30' family room, from "wall to wall."

b. Find the area at the stair entrance that does not need tile.

c. Find the area in front of the fireplace that does not need tile.

d. Find the area by the kitchen exit that does not need tile.

e. Find the total square feet of floor that is to be tiled.

CORD Applied Mathematics

EXERCISE 19:   Find the maximum diameter rod that can be inserted through a triangular hole. The triangle is an *equilateral triangle* that is 0.4 inch on a side. (Hint: An equilateral triangle is one with equal sides and equal angles.)

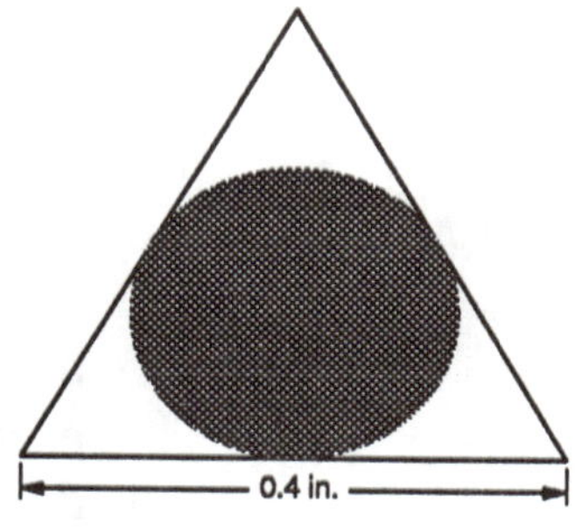

EXERCISE 20:   You want to mill a hexagonal hole punch that is 0.5 inch on each side, as shown here. What is the smallest diameter rod from which such a punch can be milled? (Hint: If you draw lines from the corners to the center of the hexagon, you will create 6 equilateral triangles.)

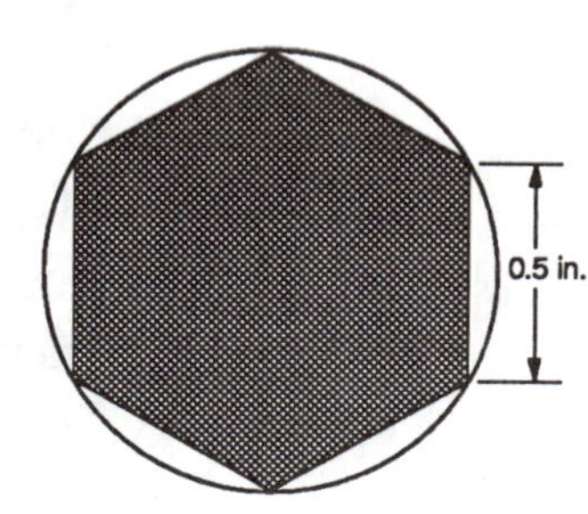

EXERCISE 21:   The rectangular heating ducts in a building need to be replaced. It would be easier to install ready-made *circular* duct material rather than reconstruct the existing *rectangular* duct. The original ductwork had a cross section of 18" × 24". The circular duct material is available in 10", 16", 24", and 36" diameters. Which of these should you select to obtain approximately the same cross-sectional area as that of the original ductwork?

EXERCISE 22:   You are installing a fence around the yard of a large house. The dimensions for the yard are shown in the drawing below. Notice the circular shape of the two corners. Calculate the perimeter of the yard so you will know how much fencing material to buy.

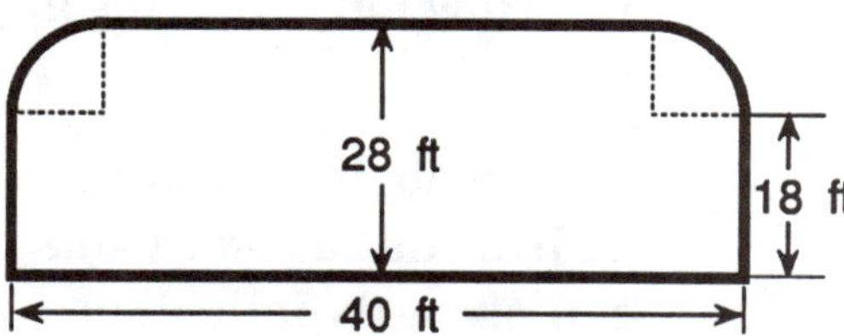

EXERCISE 23:   You have been hired to build a deck. The homeowner wants the deck shaped like a trapezoid, as shown. Since you will charge the homeowner by the square foot, you need to find the total area of the deck. The long edge of the deck will run along the 75-foot width of the

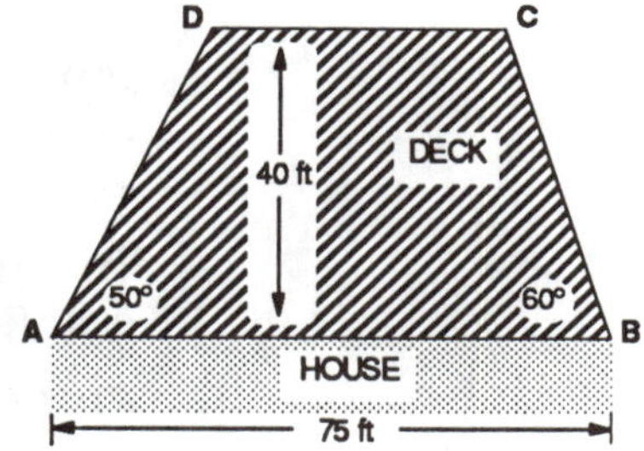

house. The distance to the back of the deck is 40 feet, as shown.
The trapezoid will form a 50° angle at one corner of the house and
a 60° angle at the other corner.

a.   Make a sketch of the deck on your paper. Divide the deck's
area into two right triangles and a rectangle. Do this by dropping
a perpendicular line from point **D** down to the long edge of the
deck, **AB**. Label the point on the long edge of the deck **E**.
Similarly drop a perpendicular line from point **C** down to the long
edge of the deck, to a point we'll label **F**. Identify the two right
triangles, △**AED** and △**BFC**, and the rectangle **CDEF**.

b.   For △**AED**, determine the length of the adjacent side, **AE**.
(Hint: Use a trig function that involves the 50° angle and the
known side opposite the angle.)

c.   For △**BFC**, perform a similar calculation to determine the
length of the adjacent side, **FB**.

d.   Use the lengths **AE** and **FB** along with the given length for
AB to determine the length **DC**.

e.   Determine the area of the deck—trapezoid **ABCD**.

   A 50-foot *vertical* pole is placed on a
hill that makes a 5° angle with the
horizontal, as shown. A guy wire is
to be attached at a point 8 feet from
the top of the pole and extended
down to the ground. The guy wire
must be attached at the ground, 10
feet from the base of the pole, up the
hill. How long will this guy wire
have to be?

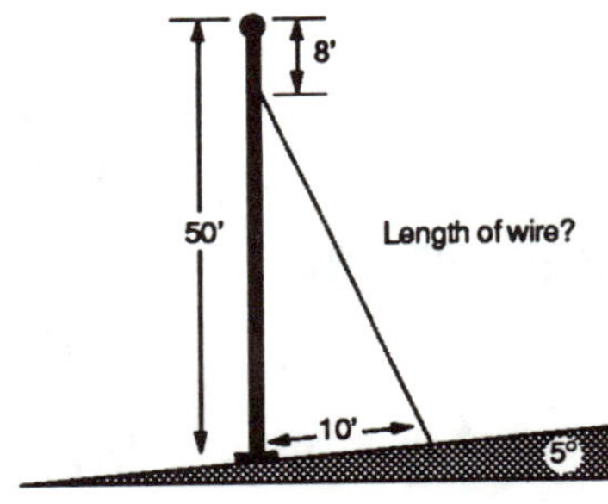

   A tooling ball is used to
confirm that a V-way has
been properly milled, as
shown here. If the depth of
the V-way is correct, the
top of the ball will be flush
with the top of the V-way.
This will confirm that the
width of the top cut is
3.500 inches. Compute the
diameter of the tooling ball ($d = 2r$) needed to confirm the V-way.

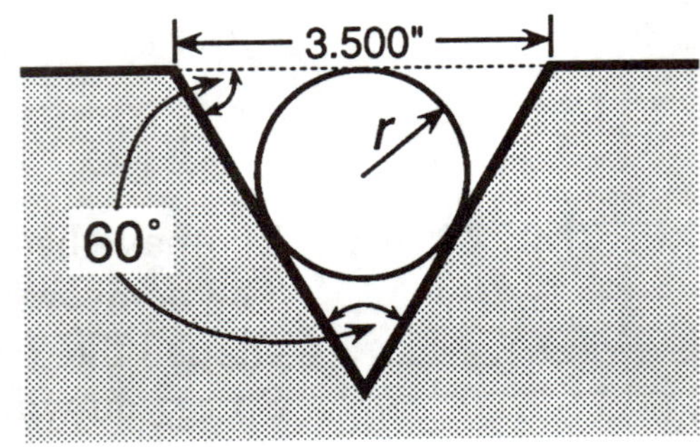

GLOSSARY

Aspect ratio for a TV tube
The ratio of height versus width for a television tube.

Chord
A line segment that joins two points on a curve.

Circle
A closed curve in two dimensions with all points on the curve at equal distance from a single, enclosed center point.

Circumference
The distance around a circle.

Clearance
The distance between mating parts in a device—such as a piston and the walls of the cylinder by which the piston moves.

Cosine
For a specified acute angle of a right triangle, the cosine of the angle is equal to the ratio of lengths of the adjacent leg to the hypotenuse.

Diameter
A line segment beginning on one side of a circle, passing through the center, and ending on the other side of the circle. The length of a diameter is twice the radius.

Die
Any of the various devices used for cutting out, forming, or stamping material. Generally refers to the mating part that is used in conjunction with a punch. (Also refers to one of the cubes—dice—used in games of chance.)

Equilateral triangle
A triangle with three equal sides and three equal angles. Each angle equals 60 degrees.

Hypotenuse
The longest side of a right triangle, always located opposite the right angle.

Pie chart
A circle chart with radii that divide the circle into sectors proportional in angle and area to the relative size of the quantities represented.

Pitch (of a screw)
The distance between two corresponding points on adjacent screw threads or gear teeth. The pitch is equal to the reciprocal of the number of threads per inch.

Punch
A tool with various cross-sectional shapes that is used to punch holes in metal.

Pythagorean theorem	A theorem that pertains to right triangles. The theorem states that the square of the length of the hypotenuse is equal to the sum of the squares of the length of each leg.
Radius	Any line segment from the center of a circle to any point on the circumference of the circle. It is equal to one-half of the diameter of a circle.
Sector	A pie-shaped piece of a circle bounded by two radii and the arc of the circle.
Segment of a circle	The area between a chord and the arc of a circle.
Sine	For a specified acute angle of a right triangle, the sine of the angle is equal to the ratio of lengths of the opposite leg to the hypotenuse.
TPI	Referring to screw threads, an abbreviation for the number of *threads per inch*.
Tangent	For a specified acute angle of a right triangle, the tangent of the angle is equal to the ratio of lengths of the opposite side to the adjacent side.
Threads	A helical or spiral ridge on a screw, nut, or bolt.
Tooling	The equipment necessary to provide a factory with the necessary machinery to manufacture products.
Trapezoid	A two-dimensional, four-sided, closed figure with two of the four sides parallel to each other.